移动互联应用开发系列

Android 项目实战
——手机安全卫士开发案例解析

王家林　王家俊　王家虎　著

电子工业出版社
Publishing House of Electronics Industry
北京·BEIJING

内 容 简 介

本书通过对一款手机安全卫士开发案例的详细解析，讲解了一个完整的 Android 实际项目的开发过程。该项目涵盖了市场上主流手机卫士的主要功能，同时，该项目也是对 Android 应用程序开发知识的综合应用。通过对案例的解析，使 Android 应用开发人员在实际开发中少走弯路，快速而轻松地积累实战项目经验。

本书适合具备一定的 Android 应用开发知识并需要提高实战开发经验的人员阅读。

未经许可，不得以任何方式复制或抄袭本书之部分或全部内容。
版权所有，侵权必究。

图书在版编目（CIP）数据

Android 项目实战：手机安全卫士开发案例解析/王家林，王家俊，王家虎著. —北京：电子工业出版社，2013.5
（移动互联应用开发系列）
ISBN 978-7-121-20084-7

Ⅰ. ①A… Ⅱ. ①王… ②王… ③王… Ⅲ. ①移动终端－应用程序－程序设计 Ⅳ. ①TN929.53

中国版本图书馆 CIP 数据核字（2013）第 064816 号

责任编辑：窦　昊
印　　刷：三河市鑫金马印装有限公司
装　　订：三河市鑫金马印装有限公司
出版发行：电子工业出版社
　　　　　北京市海淀区万寿路 173 信箱　邮编　100036
开　　本：787×1 092　1/16　印张：20.25　字数：518 千字
印　　次：2013 年 5 月第 1 次印刷
印　　数：4 000 册　定价：49.90 元

凡所购买电子工业出版社图书有缺损问题，请向购买书店调换。若书店售缺，请与本社发行部联系，联系及邮购电话：（010）88254888。
质量投诉请发邮件至 zlts@phei.com.cn，盗版侵权举报请发邮件至 dbqq@phei.com.cn。
服务热线：（010）88258888。

前　言

为什么要写这样一本书

对于一名希望能够胜任实际开发工作的 Android 应用开发人员来说，最重要的一点是什么呢？毫无疑问，就是积累丰富的项目开发经验，让自己在实际开发工作中游刃有余。

为了让那些希望能够胜任实战工作的 Android 应用开发人员少走弯路，快速而轻松地积累实战项目经验，笔者决定结合多年的项目实战开发经验，编写一本能够真正让读者学以致用的图书。

本书有何特色

为了让读者轻松地上手，本书特别设计了适合初级 Android 应用开发者的学习方式，用准确的定义总结概念，用直观的图示演示过程，用详细的注释解释代码，用简要的语言概括知识点。

① 项目模块介绍

简洁、清晰是其显著特点，一般放在每一个模块的开始部分，让读者对每一个模块都有一个清晰、全局的认识。

② 代码解析

将代码中的关键代码行逐一解释，有助于读者掌握相关概念和知识。

③ 运行结果

对每个模块均给出运行结果和对应图示，帮助读者更直观地理解实例代码。

④ 知识点小结

每完成一个模块，都会对本模块需要掌握的要点进行知识汇总。

本书适合哪些读者阅读

- 具备一定 Android 应用开发基础知识的学习人员；
- 了解 Android 应用开发基础知识，但还需要进一步学习的人员；
- 即将踏入（刚踏入）工作岗位、希望积累项目经验的开发人员；
- 其他编程爱好者。

项目介绍

本书通过对一款手机安全卫士开发案例的详细解析，讲解一个完整的 Android 实际项目的开发过程。该项目涵盖了市场上主流手机卫士的主要功能，同时，该项目也是对 Android 应用程序开发知识的综合应用。项目具体实现了九大功能：手机防盗、通信卫士、软件管理、进程管理、流量统计、手机杀毒、系统优化、高级工具、设置中心。

手机防盗：根据手机 sim 卡的变更来判断手机是否被盗，如果 sim 卡发生变更，程序会根据事先约定好的协议向绑定的安全号码发送一些信息（例如，"sim card changed"等信息），我们可以通过安全号码给手机发送一些指令（例如，远程锁频、销毁数据、播放报警音乐、获取当前手机的经纬度信息），也可以防止应用程序被卸载（例如，将"手机防盗"字样修改为"MP3"）。

通信卫士：来电黑名单管理（电话拦截、短信拦截、电话和短信拦截）。

软件管理：动态计算出当前手机的可用内存、Sdcard 可用内存。同时，以列表的形式显示手机中所有的应用程序（图标、名称、版本号），单击每一个应用程序时，弹出卸载、启动、分享的选项。

进程管理：列出手机中当前正在运行的所有进程，可以将其分为系统进程和用户进程，可以实现对进程的一键清理。

流量管理：显示手机中每个具有 Internet 权限应用程序的上传与下载所产生的流量及流量总和。

手机杀毒：根据程序特征码来识别、查杀病毒。

系统优化：清理应用程序在手机中产生的缓存文件。

高级工具：手机号码归属地查询、常用号码查询、程序锁。

设置中心：是否开启程序的自动更新、是否开启来电归属地的显示、更改归属地的显示风格、更改归属地在屏幕上的显示位置、是否开启黑名单服务、是否开启程序锁服务。

项目的实现流程及说明

项目的基本实现流程是：Splash 界面开发→手机防盗→设置中心→通信卫士→软件管理→进程管理→流量管理→手机杀毒→系统优化。

说明：项目的开发环境是在 Android4.2 环境下进行的，可在 Android2.2 及 Android2.2 以上的版本上运行。在项目实现的过程中，模块与模块之间存在一定的联系，所以在开发的过程中，有时需要进行模块间的切换（例如，在开发手机防盗功能时，需要在设置中心中设置手机防盗是否开启）。

本书的源代码及部分代码文本文件可在电子工业出版社官网"在线资源"中下载，也可以关注微博@电子社通信分社，进入微盘下载相关源码。

笔者

目 录 CONTENTS

第 1 章 项目简介与 Splash 界面开发 ... 1
1.1 创建应用 ... 1
- 1.1.1 Splash 界面的 UI 开发 ... 3
- 1.1.2 Splash 界面加载时的具体流程 ... 5
- 1.1.3 服务器端的搭建 ... 6
- 1.1.4 连接服务器获取更新信息 ... 6
- 1.1.5 下载服务端的 apk 文件 ... 14
- 1.1.6 替换安装下载后的 apk ... 18
- 1.1.7 apk 的替换安装细节 ... 23

1.2 程序主界面的 UI 设计 ... 26
1.3 关闭自动更新 ... 34

第 2 章 手机防盗模块的设计 ... 40
2.1 手机防盗的功能介绍 ... 40
2.2 手机防盗的细节 ... 49
2.3 实现手机防盗中的设置向导 UI ... 54
2.4 获取联系人的数据与完成设置向导逻辑 ... 74
2.5 实现手机防盗指令 ... 82

第 3 章 高级工具模块的设计 ... 95
3.1 号码归属地数据库的优化和复制 ... 95
3.2 号码归属地查询 ... 98
3.3 显示来电与外拨电话的号码归属地 ... 110
3.4 更改归属地的显示风格 ... 125
3.5 更改归属地的显示位置 ... 132
3.6 使用 ExpandableListView 实现常用号码的查询 ... 148
3.7 程序锁的设计和 UI ... 163
- 3.7.1 程序锁的实现 ... 164

3.7.2 程序锁中的 bug 解决方案 ································· 189

第 4 章 通信卫士模块的设计 ································· 204
4.1 通信卫士的功能介绍与 UI 设计 ································· 204
4.2 黑名单号码的添加与修改 ································· 221
4.3 黑名单号码对短信和电话的拦截 ································· 223
4.4 黑名单号码对电话的拦截 ································· 225
4.5 采用内容观察者删除呼叫记录 ································· 234

第 5 章 其他模块的设计 ································· 238
5.1 软件管理模块设计 ································· 238
5.1.1 软件管理器之分类显示应用程序 ································· 238
5.1.2 使用 PopupWindow 显示程序的启动、分享、卸载 ································· 249
5.1.3 实现程序的卸载、启动、分享功能 ································· 252
5.2 进程管理器的设计 ································· 254
5.2.1 进程管理器的实现 ································· 254
5.2.2 使用自定义吐司显示清理结果 ································· 264
5.3 流量管理模块的设计 ································· 266
5.3.1 流量统计的原理 ································· 266
5.3.2 流量统计的实现 ································· 271
5.4 手机杀毒模块的设计 ································· 282
5.4.1 病毒查杀的原理 ································· 282
5.4.2 手机杀毒的具体实现方法 ································· 283
5.5 系统优化的功能介绍与 UI 设计 ································· 296
5.5.1 采用反射技术来调用系统隐藏的 API ································· 297
5.5.2 系统优化的具体实现 ································· 302

第1章 项目简介与 Splash 界面开发

1.1 创建应用

创建工程文件，应用名称为"手机安全卫士"，工程名为"mobilesafe"，应用包名为"com.guoshisp.mobilesafe"，紧接着将要创建的 Activity 命名为 SplashActivity，该 Activity 用于向用户展现一个 Splash 界面。"Splash"在英文中被译为飞洒、飞溅。

Splash 界面的主要作用：

（1）展现产品的 LOGO，提升产品的知名度。

（2）初始化的操作（初始化数据库、文件的复制、配置的读取）。

（3）根据系统的时间或者日期做出相应的判断来加载不同的 Splash 界面（例如，QQ 的登录界面），提升用户体验。

（4）连接服务器，检查获取更新信息，提示用户升级。在我们的项目中是用于连接服务器，检查版本是否需要更新下载，以及初始化数据库。

新建 Android 项目 mobilesafe，如图 1-1 所示。将 MainActivity 改名为 SplashActivity，如

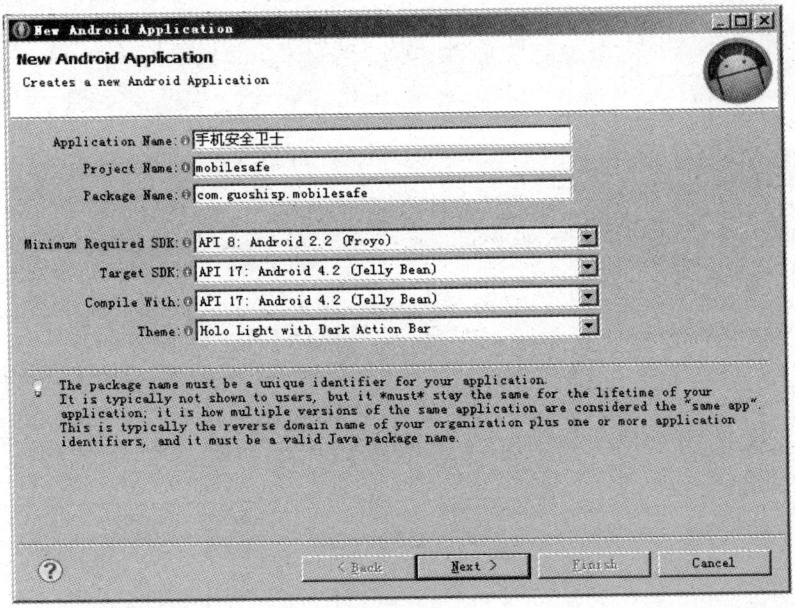

图 1-1

图 1-2 所示。在 res 目录下新建一个 drawable 目录，并将"appicon.png"图片复制到 drawable 文件中（创建该文件的目的在于：原本我们是需要提供三套图片资源来进行屏幕的适配，如果创建了该文件，只需要一套图片资源即可）作为应用的图标，如图 1-3 所示。

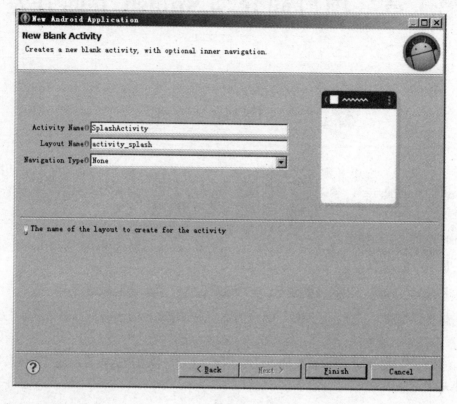

图 1-2

接下来在清单文件中修改应用的图标，如图 1-4 所示。

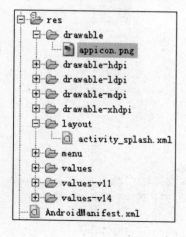

图 1-3

图 1-4

1.1.1 Splash 界面的 UI 开发

在 activity_splash.xml 中实现 Splash 界面的 UI 展示。Splash 界面的效果是：将图片 logo2 复制到 drawable 下作为界面的背景图片，在界面的右上角展示当前应用的版本号信息，在界面的中央显示一个 ProgressBar。另外该界面在屏幕上显示时无标题栏且全屏显示（这里是在 SplashActivity 中实现该设置的，当然也可以在清单文件中配置）。效果图如图 1-5 所示。

图 1-5

图 1-5 对应的 activity_splash.xml 的 UI 界面代码如下：

```xml
<RelativeLayout xmlns:android="http://schemas.android.com/apk/res/android"
    xmlns:tools="http://schemas.android.com/tools"
    android:layout_width="match_parent"
    android:layout_height="match_parent"
    tools:context=".SplashActivity"
    android:id="@+id/rl_splash"
    android:background="@drawable/logo2" >
    <TextView
        android:id="@+id/tv_splash_version"
        android:layout_width="wrap_content"
        android:layout_height="wrap_content"
        android:layout_alignParentRight="true"
        android:text="版本号:"
        android:textColor="#0aff00"
        android:textSize="20sp" />
    <ProgressBar
        android:layout_centerHorizontal="true"
        android:layout_alignParentBottom="true"
        android:layout_marginBottom="110dip"
```

```xml
        android:layout_width="wrap_content"
        android:layout_height="wrap_content" />
</RelativeLayout>
```

图 1-5 对应的 SplashActivity 业务代码如下：

```java
package com.guoshisp.mobilesafe;
import android.app.Activity;
import android.content.pm.PackageInfo;
import android.content.pm.PackageManager;
import android.os.Bundle;
import android.view.Window;
import android.view.WindowManager;
import android.widget.RelativeLayout;
import android.widget.TextView;
public class SplashActivity extends Activity {
    private TextView tv_splash_version;
    @Override
    protected void onCreate(Bundle savedInstanceState) {
        super.onCreate(savedInstanceState);
        //设置为无标题栏
        requestWindowFeature(Window.FEATURE_NO_TITLE);
        //设置为全屏模式
        getWindow().setFlags(WindowManager.LayoutParams.FLAG_FULLSCREEN,
                WindowManager.LayoutParams.FLAG_FULLSCREEN);
        setContentView(R.layout.activity_splash);
            tv_splash_version = (TextView) findViewById(R.id.tv_splash_
                                    version);
        tv_splash_version.setText("版本号:" + getVersion());
    }
    /**
     * 获取当前应用程序的版本号
     *
     * @return
     */
    private String getVersion() {
    //得到系统的包管理器，已经得到了 apk 的面向对象的包装
        PackageManager pm = this.getPackageManager();
        try {
    //参数一：当前应用程序的包名；参数二：可选的附加消息，这里用不到，可以定义为 0
            PackageInfo info = pm.getPackageInfo(getPackageName(), 0);
            //返回当前应用程序的版本号
            return info.versionName;
        } catch (Exception e) {//包名未找到异常,理论上,该异常不可能会发生
            e.printStackTrace();
            return "";
        }
    }
}
```

第 1 章 项目简介与 Splash 界面开发

代码解析：

通过调用 getVersion()方法来获取应用程序的版本号信息。版本号存在于我们的 apk 对应的清单文件中（直接解压 apk 后，即可看到对应的清单文件），版本号是 manifest 节点中的 android:versionName="1.0"。当一个应用程序被装到手机后，该 apk 被复制到手机的 data/app 目录下（也就是系统中），如图 1-6 所示。所以想得到版本号，就需要先得到与系统相关的服务（这里我们通过上、下文得到 PackageManager 系统服务），通过服务就可以得到 apk 中的信息。

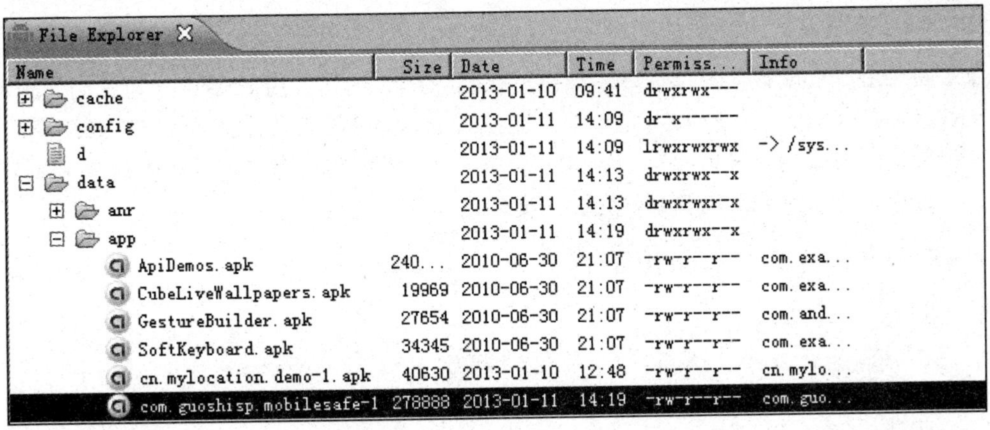

图 1-6

1.1.2 Splash 界面加载时的具体流程

接下来，我们来处理 Splash 界面在加载进来时的具体实现方式：

（1）为 Splash 界面做一个淡进（由暗变明）的动画效果，动画播放时间为 2 秒。

（2）在 Splash 界面加载时，连接服务器检查软件是否需要更新（后面还需要实现对数据库的复制），其流程图如图 1-7 所示。

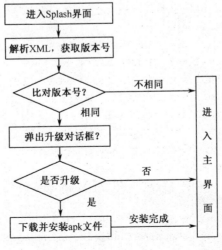

图 1-7

1.1.3 服务器端的搭建

当解析服务端的配置文件时,服务端的配置文件应当包含以下三条信息:

- 版本号,即 versionName。如果版本号为 2.0,请不要将其写为 2;
- 版本升级描述信息,即 description,用于提示用户升级的信息;
- 新版本的 apk 的下载路径。

我们将服务端的配置信息以 XML 文件格式进行存储。这里使用 Tomcat 作为我们的服务器。配置文件存放在 Tomcat 的 ROOT 目录下,文件为 info.xml(在编写该文件时需要注意一点:编写文件时采用的编码方式需要与 encoding 保持一致,否则在使用浏览器解析时会出现字符解析错误,这里采用的是 utf-8 的编码格式),文件内容如下:

```xml
<?xml version="1.0" encoding="utf-8"?>
<info>
    <version>1.0</version>
    <description>亲,有新版本了,赶紧来下载吧!</description>
    <apkurl>http://192.168.0.4:8080/mobilesafe.apk</apkurl>
</info>
```

同时将最新的 apk 复制到 Tomcat 的 ROOT 目录下。如果服务端的配置信息中的 version 为 2.0,那么所对应的 apk 中的 versionCode="2",versionName="2.0"。如果 versionName="1.0",而且本地的 apk 中也是 1.0,即使更新安装后,下次进入时还会提示升级。这是因为本地的 versionName 和服务端的 version 的值不相同。

1.1.4 连接服务器获取更新信息

1. SplashActivity.java 对应的核心业务代码

```java
Public class SplashActivity extends Activity {
private UpdateInfo info;
private static final int GET_INFO_SUCCESS = 10;
private static final int SERVER_ERROR = 11;
private static final int SERVER_URL_ERROR = 12;
private static final int PROTOCOL_ERROR = 13;
private static final int IO_ERROR = 14;
private static final int XML_PARSE_ERROR = 15;
protected static final String TAG = "SplashActivity";
private long startTime;
private RelativeLayout rl_splash;
private long endTime;
private Handler handler = new Handler() {
    public void handleMessage(android.os.Message msg) {
        switch (msg.what) {
```

```java
            case XML_PARSE_ERROR:
                Toast.makeText(getApplicationContext(), "xml解析错误",1).show();
                break;
            case IO_ERROR:
                Toast.makeText(getApplicationContext(), "I/O错误", 1).show();
                break;
            case PROTOCOL_ERROR:
                Toast.makeText(getApplicationContext(), "协议不支持", 1).show();
                break;
            case SERVER_URL_ERROR:
                Toast.makeText(getApplicationContext(), "服务器路径不正确",
                        1).show();
                break;
            case SERVER_ERROR:
                Toast.makeText(getApplicationContext(), "服务器内部异常",
                        1).show();
                break;
            case GET_INFO_SUCCESS:
                String serverversion = info.getVersion();
                String currentversion = getVersion();
                if (currentversion.equals(serverversion)) {
                    Log.i(TAG, "版本号相同进入主界面");
                } else {
                    Log.i(TAG, "版本号不相同,升级对话框");
                    showUpdateDialog();//显示升级对话框
                }
                break;
            }
        };
    };
    @Override
    public void onCreate(Bundle savedInstanceState) {
        super.onCreate(savedInstanceState);
        //设置为无标题栏
        requestWindowFeature(Window.FEATURE_NO_TITLE);
        //设置为全屏模式
        getWindow().setFlags(WindowManager.LayoutParams.FLAG_FULLSCREEN,
                WindowManager.LayoutParams.FLAG_FULLSCREEN);
        setContentView(R.layout.activity_splash);
        rl_splash = (RelativeLayout) findViewById(R.id.rl_splash);
        tv_splash_version = (TextView) findViewById(R.id.tv_splash_version);
        tv_splash_version.setText("版本号:" + getVersion());
        AlphaAnimation aa = new AlphaAnimation(0.3f, 1.0f);
        aa.setDuration(2000);
        rl_splash.startAnimation(aa);
        //连接服务器获取服务器上的配置信息
```

```java
        new Thread(new CheckVersionTask() {
        }.start();
    }
    /**
     * 连网检查应用的版本号与服务端上的版本号是否相同*
     */
    private class CheckVersionTask implements Runnable {
        public void run() {
            startTime = System.currentTimeMillis();
            Message msg = Message.obtain();
            try {
                //获取服务端的配置信息的连接地址
                String serverurl = getResources().getString(R.string.serverurl);
                URL url = new URL(serverurl);
                HttpURLConnection conn = (HttpURLConnection) url
                        .openConnection();
                conn.setRequestMethod("GET");//设置请求方式
                conn.setConnectTimeout(5000);
                int code = conn.getResponseCode();//获取响应码
                if (code == 200) {//响应码为 200 时,表示与服务端连接成功
                    InputStream is = conn.getInputStream();
                    info = UpdateInfoParser.getUpdateInfo(is);
                    endTime = System.currentTimeMillis();
                    long resulttime = endTime - startTime;
                    if (resulttime < 2000) {
                        try {
                            Thread.sleep(2000 - resulttime);
                        } catch (InterruptedException e) {
                            e.printStackTrace();
                        }
                    }
                    msg.what = GET_INFO_SUCCESS;
                    handler.sendMessage(msg);
                } else {
                    //服务器状态错误
                    msg.what = SERVER_ERROR;
                    handler.sendMessage(msg);
                    endTime = System.currentTimeMillis();
                    long resulttime = endTime - startTime;
                    if (resulttime < 2000) {
                        try {
                            Thread.sleep(2000 - resulttime);
                        } catch (InterruptedException e) {
                            e.printStackTrace();
                        }
                    }
```

```java
                }
            } catch (MalformedURLException e) {
                e.printStackTrace();
                msg.what = SERVER_URL_ERROR;
                handler.sendMessage(msg);
            } catch (ProtocolException e) {
                msg.what = PROTOCOL_ERROR;
                handler.sendMessage(msg);
                e.printStackTrace();
            } catch (IOException e) {
                msg.what = IO_ERROR;
                handler.sendMessage(msg);
                e.printStackTrace();
            } catch (XmlPullParserException e) {
                msg.what = XML_PARSE_ERROR;
                handler.sendMessage(msg);
                e.printStackTrace();
            }
        }
    }
    /**
     * 显示升级提示的对话框
     */
    protected void showUpdateDialog() {
        //创建对话框的构造器
        AlertDialog.Builder builder = new Builder(this);
        //设置对话框提示标题左边的提示图标
        builder.setIcon(getResources().getDrawable(R.drawable.notification));
        //设置对话框的标题
        builder.setTitle("升级提示");
        //设置对话框的提示内容
        builder.setMessage(info.getDescription());
        //设置升级按钮
        builder.setPositiveButton("升级", new OnClickListener() {
        //设置取消按钮
        public void onClick(DialogInterface dialog, int which) {
        });
        builder.setNegativeButton("取消", new OnClickListener() {
        });
        //创建并显示对话框
        builder.create().show();
    }
```

代码解析：

（1）rl_splash = (RelativeLayout) findViewById(R.id.rl_splash)找到 Splash 界面的根节点，为其播放动画做准备。

（2）为 Splash 界面播放一个动画。

其中，AlphaAnimation aa = new AlphaAnimation(0.3f, 1.0f)设置一个透明度由 0.3f～1.0f 的淡入的动画效果。0.0f 表示完全透明，1.0f 表示正常显示效果。

aa.setDuration(2000)为设置动画的执行时间（0.3f～1.0f），单位为毫秒。

rl_splash.startAnimation(aa)为启动动画。

（3）new Thread(new CheckVersionTask()) {
　　}.start()

由于连网的过程一般是一个耗时的操作，为了避免出现 ANR 异常，我们在主线程中开启一个子线程用于连网核对版本号信息。此时我们还应当在清单文件中配置网络权限信息：
<uses-permission android:name="android.permission.INTERNET"/>。

（4）startTime = System.currentTimeMillis()：用于记录子线程开始执行的时间。

（5）Message msg = Message.obtain()：得到一个向主线程发送消息的消息对象。

（6）String serverurl =getResources().getString(R.string.serverurl)：得到访问服务端配置信息（即服务端的 info.xml 文件）的 URL 地址。

该 URL 地址存放在 values 文件夹下的 config.xml 文件中（将请求服务端的 URL 存放在这里的目的在于：当这个 URL 需要变更时，不需要在代码中进行修改，只需要修改这个配置文件即可实现，降低了开发的成本），代配置信息如下：

```
<?xml version="1.0" encoding="utf-8"?>
<resources>
    <string name="serverurl">http://192.168.0.4:8080/info.xml</string>
</resources>
```

（7）取得与服务器的连接。

URL url = new URL(serverurl)为使用 serverurl 建立 URL 类对象 HttpURLConnection conn = (HttpURLConnection) url.openConnection()通过 URL 对象打开路径的链接，实际上这个类内部是使用 HttpURLConnection 来实现的，得到 urlConnection 对象 conn.setConnectTimeout(5000)，在 conn 对象中调用 setConnectTimeout()方法，设置链接的超时时间。之所以要设置这个超时时间，是因为：如果请求时间比较长，比如要等 30 秒，那么这个程序就处在一个阻塞过程中，而 Android 组件也有一个阻塞时间，笔者没有精确计算过，但是建议最好不要超过 6 秒，超过 6 秒时，Android 系统就会自动判断，只要超过了它的阻塞时间，不管你的程序是否有错，都会被系统回收。有些请求工作时间比较长的，千万不能在主线程中处理，主线程一旦被阻塞了，一定会被回收。碰到这种情况我们最好抛开主线程去做，而不要在主线程中处理。如果不是开发 Android 应用，而是 J2SE 应用，不设也无所谓，因为它没有超时这个说法。

（8）得到数据。

InputStream is = conn.getInputStream()，用 conn 对象调用 getInputStream()方法得到服务端

配置信息文件的输入流。

info = UpdateInfoParser.getUpdateInfo(is)将输入流传入并通过对 XML 的解析获取 XML 中的数据，并将获取的数据存放在 info 对象（UpdateInfo）中返回来。

（9）endTime = System.currentTimeMillis()，其中 endTime 表示的是从服务端返回数据时的当前时间。

（10）long resulttime = endTime – startTime，其中 resulttime 表示的是向服务端发送请求并得到对应的数据后的所用时间。

（11）if (resulttime < 2000) {
　　　　　　　try {
　　　　　　　　　Thread.sleep(2000 - resulttime);
　　　　　　　} catch (InterruptedException e) {
　　　　　　　　　e.printStackTrace();
　　　　　　　}
　　　}

（12）private Handler handler = new Handler() {
　　　public void handleMessage(android.os.Message msg) {}
　　　}

该对象是为了接收子线程发送过来的消息（主线程与子线程进行通信），将子线程发送过来的消息在 handleMessage(android.os.Message msg)方法中进行处理。

（13）msg.what = GET_INFO_SUCCESS：为该 msg 做一个标记，这样，在 Handler 的 handleMessage(android.os.Message msg)的 Switch 可以获取到该标记，以便识别是哪个消息。

（14）handler.sendMessage(msg)：通过 handler 对象向主线程中发送消息，然后在 Handler 的 handleMessage(android.os.Message msg)方法中可以处理该消息。

Splash 界面播放动画的时间设置为 2 秒，在播放这个 Splash 动画的过程中也进行连网检查更新数据的操作，连网检查更新数据结束后，如果不需要更新数据，那么将直接跳转到主界面（MainActivity）。在网络较好的情况下，连网时间不会超过 2 秒（可能不到 0.5 秒），这时如果没有（11）的代码就会直接跳转到主界面，当然，如果连网检查更新的过程超过了 2 秒，就需要等到检查更新后进入主界面。

2．解析 XML 的业务方法

在解析服务端的 info.xml 文件时，需要创建出一个解析 XML 的业务方法。将该业务方法放在 "com.guoshisp.mobilesafe.engine" 包下的 UpdateInfoParser 类中。具体代码如下：

```
package com.guoshisp.mobilesafe.engine;
import java.io.IOException;
import java.io.InputStream;
```

```java
import org.xmlpull.v1.XmlPullParser;
import org.xmlpull.v1.XmlPullParserException;
import android.util.Xml;
import com.guoshisp.mobilesafe.domain.UpdateInfo;
/**
 *
 * 解析 XML 数据
 *
 */
public class UpdateInfoParser {
    /**
     * @param is XML 文件的输入流
     * @return updateinfo 的对象
     * @throws XmlPullParserException
     * @throws IOException
     */
    public static UpdateInfo getUpdateInfo(InputStream is)
            throws XmlPullParserException, IOException {
        //获得一个 Pull 解析的实例
        XmlPullParser parser = Xml.newPullParser();
        //将要解析的文件流传入
        parser.setInput(is, "UTF-8");
        //创建 UpdateInfo 实例，用于存放解析得到的 XML 中的数据，最终将该对象返回
        UpdateInfo info = new UpdateInfo();
        //获取当前触发的事件类型
        int type = parser.getEventType();
        //使用 while 循环，如果获得的事件码是文档结束，那么就结束解析
        while (type != XmlPullParser.END_DOCUMENT) {
            if (type == XmlPullParser.START_TAG) {//开始元素
                if ("version".equals(parser.getName())) {//判断当前元素是否
                    是读者需要检索的元素，下同
                    //因为内容也相当于一个节点，所以获取内容时需要调用 parser 对象的 nextText()
                    方法才可以得到内容
                    String version = parser.nextText();
                    info.setVersion(version);
                } else if ("description".equals(parser.getName())) {
                    String description = parser.nextText();
                    info.setDescription(description);
                } else if ("apkurl".equals(parser.getName())) {
                    String apkurl = parser.nextText();
                    info.setApkurl(apkurl);
                }
            }
```

```
            type = parser.next();
        }
        return info;
    }
}
```

代码解析：

（1）int type = parser.getEventType()，这是 Pull 解析器的第一个事件。读者可以看到，这个方法的返回值是 int 类型，Pull 解析器返回的是一个数字，类似于一个信号。那么这些信号都代表什么意思呢？Pull 解析器已经定义了这五个常量，而且对于事件，仅仅只有这五个，如下：

- XmlPullParser.START_DOCUMENT——开始解析；
- XmlPullParser.START_TAG——开始元素；
- XmlPullParser.TEXT——解析文本；
- XmlPullParser.END_TAG——结束元素；
- XmlPullParser.END_DOCUMENT——结束解析。

（2）parser.getEventType()触发了第一个事件，根据 XML 的语法，也就是从它开始来解析文档。那么，怎样触发下一个事件呢？要通过 parser 中最重要的方法：parser.next()。

注意：该方法是有返回值的，在 Pull 触发下一个事件的同时，我们也获得该事件的"信号"，通过获得的信号进行 switch 操作。

3．实体数据封装

当解析完 XML 文件后，将解析出来的实体数据封装在"com.guoshisp.mobilesafe. domain"包下的 UpdateInfo 类中。具体代码如下：

```
package com.guoshisp.mobilesafe.domain;
public class UpdateInfo {
    private String version;//服务端的版本号
    private String description;//服务端的升级提示
    private String apkurl;//服务端的 apk 下载地址
    public String getVersion() {
        return version;
    }
    public void setVersion(String version) {
        this.version = version;
    }
    public String getDescription() {
        return description;
    }
    public void setDescription(String description) {
        this.description = description;
    }
```

```
    public String getApkurl() {
        return apkurl;
    }
    public void setApkurl(String apkurl) {
        this.apkurl = apkurl;
    }
}
```

测试运行：服务端的版本配置信息中的 version 设置为 2.0（本地的 apk 的版本号是 1.0，服务端的 apk 版本号修改为 2.0），运行效果如图 1-8 所示，日志打印如图 1-9 所示。

图 1-8   图 1-9

1.1.5　下载服务端的 apk 文件

当程序连接服务器检测到新版本并弹出升级对话框的提示时，我们可以单击"下载"按钮将服务器端的最新 apk 下载到本地的 Sdcard 上。单击"取消"按钮时，直接进入当前版本程序的主界面。显示升级提示对话框的核心代码如下：

```
/**
 * 显示升级提示的对话框
 */
protected void showUpdateDialog() {
        //创建对话框的构造器
        AlertDialog.Builder builder = new Builder(this);
        //设置对话框的提示内容
builder.setIcon(getResources().getDrawable(R.drawable.notification));
        //设置升级标题
        builder.setTitle("升级提示");
        //设置升级提示内容
        builder.setMessage(info.getDescription());
        //创建下载进度条
        pd = new ProgressDialog(SplashActivity.this);
        //设置进度条在显示时的提示消息
        pd.setMessage("正在下载");
```

```java
//指定显示下载进度条为水平形状
pd.setProgressStyle(ProgressDialog.STYLE_HORIZONTAL);
//设置升级按钮
builder.setPositiveButton("升级", new OnClickListener() {
    public void onClick(DialogInterface dialog, int which) {
        Log.i(TAG, "升级,下载" + info.getApkurl());
        //判断 Sdcard 是否存在
        if (Environment.MEDIA_MOUNTED.equals(Environment
                .getExternalStorageState())) {
            pd.show();//显示下载进度条
            //开启子线程下载 apk
            new Thread() {
                public void run() {
                    //获取服务端新版本 apk 的下载地址
                    String path = info.getApkurl();
                    //获取最新 apk 的文件名
                    String filename = DownLoadUtil.getFilename(path);
                    //在 Sdcard 上创建一个文件
                    File file = new File(Environment
                            .getExternalStorageDirectory(), filename);
                    //得到下载后的 apk 文件
                    file = DownLoadUtil.getFile(path,
                            file.getAbsolutePath(), pd);
                    if (file != null) {
                        //向主线程发送下载成功的消息
                        Message msg = Message.obtain();
                        msg.what = DOWNLOAD_SUCCESS;
                        msg.obj = file;
                        handler.sendMessage(msg);
                    } else {
                        //向主线程发送下载失败的消息
                        Message msg = Message.obtain();
                        msg.what = DOWNLOAD_ERROR;
                        handler.sendMessage(msg);
                    }
                    pd.dismiss();//下载结束后,将下载的进度条关闭
                };
            }.start();
        } else {
            Toast.makeText(getApplicationContext(), "sd 卡不可用",
                    1).show();
            loadMainUI();//进入程序主界面
        }
    }
});
builder.setNegativeButton("取消", new OnClickListener() {
```

```
            public void onClick(DialogInterface dialog, int which) {
                loadMainUI();
            }
        });
        builder.create().show();
    }
```

说明：由于该方法在主线程中执行，而下载文件又是一个比较耗时的操作，所以我们开启一个新的线程来执行该任务。而且，在下载完成后，我们必须调用 pd 对象的 dismiss()方法，否则该进度条不会自动消失。

下载 apk 的工具类位于"com.guoshisp.mobilesafe.utils"包下的 ownLoadUtil，该类的具体实现代码如下：

```
package com.guoshisp.mobilesafe.utils;
import java.io.File;
import java.io.FileOutputStream;
import java.io.InputStream;
import java.net.HttpURLConnection;
import java.net.MalformedURLException;
import java.net.URL;
import android.app.ProgressDialog;
import android.content.Context;
import android.os.Environment;
/**
 * 下载的工具类：下载文件的路径；下载文件后保存的路径；关心进度条；上、下文
 */
public class DownLoadUtil {
    /**
     * 下载一个文件
     *
     * @param urlpath
     * 路径
     * @param filepath
     * 保存到本地的文件路径
     * @param pd
     * 进度条对话框
     * @return 下载后的 apk
     */
    public static File getFile(String urlpath, String filepath,
            ProgressDialog pd) {
        try {
            URL url = new URL(urlpath);
            File file = new File(filepath);
            FileOutputStream fos = new FileOutputStream(file);
            HttpURLConnection conn = (HttpURLConnection) url.openConnection();
```

```java
            //下载的请求是 GET 方式，conn 的默认方式也是 GET 请求
            conn.setRequestMethod("GET");
            //服务端的响应时间
            conn.setConnectTimeout(5000);
            //获取服务端的文件总长度
            int max = conn.getContentLength();
            //将进度条的最大值设置为要下载的文件的总长度
            pd.setMax(max);
            //获取要下载的 apk 的文件的输入流
            InputStream is = conn.getInputStream();
            //设置一个缓存区
            byte[] buffer = new byte[1024];
            int len = 0;
            int process = 0;
            while ((len = is.read(buffer)) != -1) {
                fos.write(buffer, 0, len);
                //每读取一次输入流，就刷新一次下载进度
                process+=len;
                pd.setProgress(process);
                //设置睡眠时间，便于观察下载进度
                Thread.sleep(30);
            }
            //刷新缓存数据到文件中
            fos.flush();
            //关流
            fos.close();
            is.close();
            return file;
        } catch (Exception e) {
            e.printStackTrace();
            return null;
        }
    }
    /**
     * 获取一个路径中的文件名。例如：mobilesafe.apk
     *
     * @param urlpath
     * @return
     */
    public static String getFilename(String urlpath) {
        return urlpath
                .substring(urlpath.lastIndexOf("/") + 1, urlpath.length());
    }
}
```

由于涉及对 Sdcard 的操作，所以我们需要在清单文件中配置相应的权限：

```xml
<uses-permission
    android:name="android.permission.WRITE_EXTERNAL_STORAGE" />
<uses-permission
    android:name="android.permission.MOUNT_UNMOUNT_ FILESYSTEMS" />
```

当再次运行并单击"升级"按钮时,界面效果如图 1-10 所示。

图 1-10

1.1.6 替换安装下载后的 apk

当新版本的 apk 下载到手机的 Sdcard 目录下时,此时应该立即安装这个新版本的 apk。通过查看 Android 的源代码 dir\JB\packages\apps\PackageInstaller 目录下的 AndroidManifest.xml 清单文件可以看到这样一个意图过滤器:

```xml
<intent-filter>
    <action android:name="android.intent.action.VIEW" />
    <category android:name="android.intent.category.DEFAULT" />
    <data android:scheme="content" />
    <data android:scheme="file" />
    <data android:mimeType="application/vnd.android.package-archive" />
</intent-filter>
```

所以,根据此意图过滤器,我们可以写出对应的方法来激活这个意图:

```java
/**
 * 安装一个 apk 文件
 *
 * @param file 要安装的完整文件名
 */
protected void installApk(File file) {
    //隐式意图
```

```
            Intent intent = new Intent();
            intent.setAction("android.intent.action.VIEW");//设置意图的动作
            intent.addCategory("android.intent.category.DEFAULT");
            //为意图添加额外的数据
            //intent.setType("application/vnd.android.package-archive");
            //intent.setData(Uri.fromFile(file));
            intent.setDataAndType(Uri.fromFile(file),
            "application/vnd.android.package-archive");//设置意图的数据与类型
            startActivity(intent);//激活该意图
        }
```

说明：这是一个隐式意图。隐式意图的激活方式：系统首先查询一个系统注册表（位于手机的 data\system\packages.xml 文件中），当查找到与之对应的数据后才将对应的组件激活，这个过程是先查询后激活，效率相对于显式意图要低一些。如果组件在不同的应用程序里面，则不能通过显式意图来激活，这时我们需要借助隐式意图。显式意图的应用场景：在当前应用程序里去激活自己的组件，直接通过指定组件名即可激活，效率较高。

此时，SplashActivity 的完整的业务代码见在线资源包中代码文本文件 1.1.6.doc。

相关代码解析：

（1）case DOWNLOAD_SUCCESS:
      Log.i(TAG, "文件下载成功");
      //得到发送过来的消息中的文件对象
      File file = (File) msg.obj;
      //将文件对象传入，执行安装方法来安装该文件
      installApk(file);
      break;

（2）intent.setDataAndType(Uri.fromFile(file),
"application/vnd.android.package-archive");//设置意图的数据与类型

调用 intent 对象的 setDataAndType(Uri.fromFile(file), "application/vnd.android.package-archive") 方法可以同时设置意图的数据与类型，这里之所以没有使用"intent.setType("application/vnd.android.package-archive")"与"intent.setData(Uri.fromFile(file))"，是因为：当执行 intent.setType("application/vnd.android. package-archive")代码时，会将前面的 intent.setData(Uri.fromFile(file))中设置的数据移除，当执行 intent.setData(Uri.fromFile(file))时，会将前面的 intent.setType("application/vnd.android. package-archive")中设置的类型移除，反之同理。而调用 intent 对象的 setDataAndType(Uri. fromFile(file), "application/vnd.android.package-archive")方法时，则可以同时设置这两者，不会出现移除现象，从而激活系统的安装意图。

当成功下载到 apk 后，程序执行到 installApk(file)方法时，会激活系统的安装意图，立即进入 apk 的安装界面来安装传入的文件，如图 1-11 所示。单击"安装"按钮后，显示出正在安装的界面，如图 1-12 所示。如果 apk 安装成功，那么当再次进入 Splash 界面时，版本号应该是 2.0，如图 1-13 所示。Log 中重要的日志信息如图 1-14 所示。

图 1-11　　　　　　　图 1-12　　　　　　　图 1-13

图 1-14

（3）AlertDialog.Builder builder = new Builder(this)。我们向构造器中传入的是 this，即 SplashActivity.this，传入的是当前 Activity 的上、下文对象。还有一种上、下文对象是 getApplicationContext()，如果传入的是 getApplicationContext()，运行程序时，会出现错误 "android.view.WindowManager$BadTokenException:Unable to add window -- token null is not for an application"，这句话的意思是：不能够添加窗体，该应用的令牌是 null。此时通过一个实验来揭开迷雾。

新建工程"AlertDialogTest"，在 Activity 中实现对有"弹出对话框"字样的按钮的监听，一旦单击该按钮，将会在当前的 Activity 中弹出一个对话框。具体业务代码如下：

```java
package com.example.alertdialogtest;
import android.app.Activity;
import android.app.AlertDialog;
import android.app.AlertDialog.Builder;
import android.content.DialogInterface;
import android.content.DialogInterface.OnClickListener;
import android.os.Bundle;
import android.view.View;
public class MainActivity extends Activity {
    @Override
    protected void onCreate(Bundle savedInstanceState) {
        super.onCreate(savedInstanceState);
        setContentView(R.layout.activity_main);
    }
    /**
```

```
 * 当Activity失去焦点时调用
 */
@Override
protected void onPause() {
System.out.println("pause");
    super.onPause();
}
/**
 * 当单击按钮时执行该方法,因为我们在对应的XML文件中的Button中设置了属性:
   android:onClick="onClick"
 * @param view
 */ public void onClick(View view ){
    //获取对话框的构造器
    AlertDialog.Builder builder = new Builder(this);
    builder.setMessage("我是对话框");
builder.setPositiveButton("确定", new OnClickListener() {
        @Override
        public void onClick(DialogInterface dialog, int which) {
            //TODO Auto-generated method stub
        }
    });
    //创建并显示对话框
    builder.create().show();
    }
}
```

其对应的activity_main布局文件为:

```
<?xml version="1.0" encoding="utf-8"?>
<LinearLayout xmlns:android="http://schemas.android.com/apk/res/android"
    android:layout_width="fill_parent"
    android:layout_height="fill_parent"
    android:orientation="vertical" >
  <Button
     android:id="@+id/button"
     android:layout_width="wrap_content"
     android:layout_height="wrap_content"
     android:text="弹出对话框" />
</LinearLayout>
```

当运行工程时,会出现一个带有"弹出对话框"按钮的界面,如图1-15所示。当单击该按钮时,会弹出一个对话框,如图1-16所示。根据Activity的生命周期我们知道,当一个Activity失去焦点时,必然会执行Activity生命周期方法中的onPause()方法,而此时,Log日志中并没有打印出所预料的"pause"信息。所以,onPause()方法肯定没有被执行。通过以上分析我们可以得出这样一个结论:Dialog窗体是Activity的一部分。然而,通过getApplicationContext()方法获取的上、下文是属于整个应用程序的,而Activity.this则是获得的当前Activity的上、下文。

如果是通过 getApplicationContext 获取上、下文，那么，系统就不知道当前的这个窗体要挂载在哪个 Activity 上（因为 getApplicationContext 是整个应用程序所共有的）。

图 1-15

图 1-16

下面再介绍一下两者的区别。

① 对于 getApplicationContext，我们可以假定它是一个父类（它属于整个应用所共有），Activity.this 可以假定为 getApplicationContext 的一个子类（当前 Activity 的上、下文），该子类中包含了一些特殊的引用（相对于父类来说，功能更加完善）。所以，一般可以用 getApplicationContext 的地方，就可以用 Activity.this 来替代。

② 生命周期上：通过 getApplicationContext 获取的上、下文对象，只要当前应用程序的进程还存在，那么该对象就一直存在；对于 Activity.this 上、下文来说，只要当前的 Activity 执行了 onDestroy()方法，这个上、下文对象也就跟着被系统回收。

③ 应用场景上：如果我们要通过一个上、下文来执行某个动作，且希望该动作一直处于"活跃"状态，那么应当考虑使用 getApplicationContext 获取的上、下文对象。例如，当使用数据库时，需要传递一个上、下文，如果传递的是 Activity.this，那么，当 Activity 执行 onDestroy()方法时，数据库就会被关闭，应用程序会出现错误。但如果使用 getApplicationContext()方法来获取上、下文对象，然后将其传递进去，那么就可以避免上面的错误。

总结：Dialog 窗体是 Activity 的一部分；一般当关乎到生命周期时，我们才会仔细分析使用哪个上、下文，一般情况下使用的是 Activity.this。

④ loadMainUI()方法用于跳转至另外一个界面 MainActivity（需要在清单文件中为其配置一下对应的信息），MainActivity 对应的业务代码如下：

```
package com.guoshisp.mobilesafe;
import android.app.Activity;
import android.os.Bundle;
public class MainActivity extends Activity {
    @Override
```

```
    protected void onCreate(Bundle savedInstanceState) {
        super.onCreate(savedInstanceState);
            setContentView(R.layout.main);
    }
}
```

其对应的 main.xml 布局文件如下：

```
<?xml version="1.0" encoding="utf-8"?>
<LinearLayout xmlns:android="http://schemas.android.com/apk/res/android"
    android:layout_width="fill_parent"
    android:layout_height="fill_parent"
    android:orientation="vertical" >
    <TextView
        android:layout_width="wrap_content"
        android:layout_height="wrap_content"
        android:layout_gravity="center_horizontal"
        android:text="我的手机卫士"
        android:textColor="#66ff00"
        android:textSize="28sp" />
</LinearLayout>
```

1.1.7　apk 的替换安装细节

当我们用鼠标右键单击工程运行时，此时 Eclipse 会采用自己默认的签名来为我们的 Android 应用签名。当然，也可以通过 Eclipse 为 Android 应用签名，然后将签名后的 apk 部署到服务器上，接着将服务端的配置信息（info.xml）中的版本号修改为 3.0。当再次执行应用时，在连网检查的过程中发现本地的版本号与服务端的 3.0 版本号不一致，此时，屏幕会弹出升级提示的对话框，当我们单击"升级"按钮下载成功后，单击"安装"按钮，在安装的过程中会出现安装失败的界面，如图 1-21 所示。

为 Android 应用签名的步骤：

用鼠标右键单击工程→左键单击"Android Tools"→左键单击"Export Signed Application Packages…"，此时弹出一个对话框，如图 1-18 所示。该对话框的目的在于确认我们要导出并签名的工程项目，默认的是我们刚才"右键工程"的项目，如果要导出并签名其他应用，我们可以在该对话框中使用鼠标左键单击"Browse…"来选择其他应用项目。如果确认无误，此时用鼠标左键单击"Next>"会进入"Keystore selection"对话框，在该对话框中，默认的是选中"Use existing keystore"。如果要为应用程序签名，需要选中"Create new keystore"来创建自己的签名。其中的"Locatiion"表示的是签名密钥存放的位置，鼠标左键单击"Browse…"选择密钥所要存放的位置，这里存放在"G:\MyApk"目录下，并将密钥命名为"myapk.keystore"。为了规范起见，文件的扩展名最好以".keystore"结尾，其中的"Password"是密钥的密码，这也是用于识别应用开发者身份的一条重要信息，密码长度不能少于 6 个字符，"Confirm"即为确认密码，如图 1-16 所示。确认密码无误后，使用鼠标左键单击"Next>"，此时弹出"Key

Creation"对话框,其中各个参数解释如下:

- Alias:key 的别名,最多只会用到前 8 个字符;
- Password:Key 的密码;
- Confirm:确认 Key 的密码;
- Validity(years):应用程序的有效期(Google 建议应用程序的有效期应当在 2033 年 10 月 22 日之后);
- First and Last Name:名字与姓氏;
- Organizational Unit:组织单位名称;
- Organization:组织名称;
- City or Locatiion:个人或机构所在的城市或区域名称;
- State or Province:城市所在的省份;
- Country Code(XX):个人或机构所在国家的国家码,如 086。

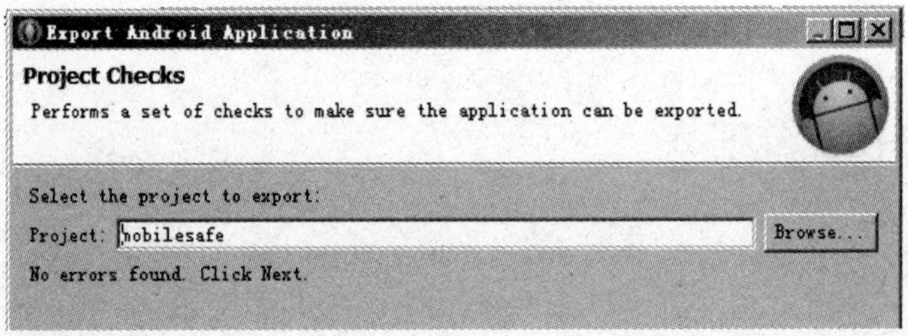

图 1-17

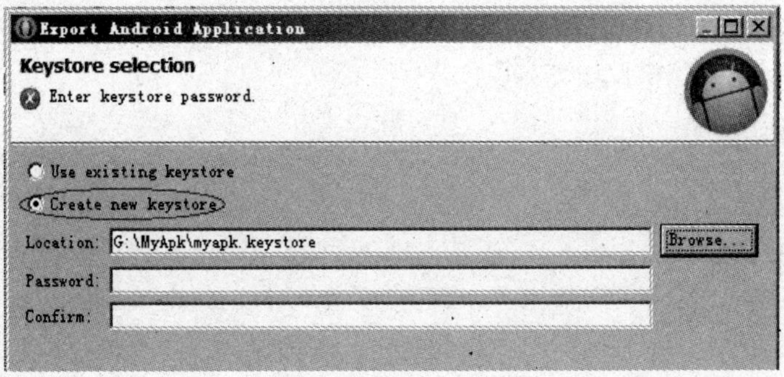

图 1-18

填写完各个参数后,如图 1-19 所示。当正确地填写完各项参数后,用鼠标左键单击"Next>"后进入"Destination and key/certificate checks"对话框,该对话框中的"Destination APK file"指定签名后的 apk 所要导出到何处。用鼠标左键单击"Browse…",选择"G:\MyApk"后界面效果图如图 1-20 所示,此时用鼠标左键单击"Next>"后即可导出签名后的 apk,这需要一个等待的过程。导出成功后,在"G:\MyApk"目录下应该生成 apk 和一个密钥,如图 1-22 所示。

第 1 章　项目简介与 Splash 界面开发

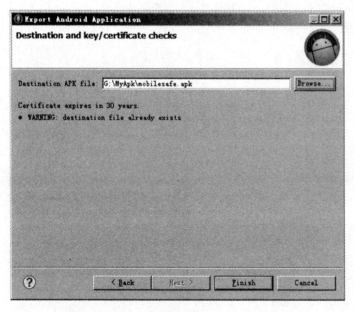

图 1-19

图 1-20　　　　　　　　　　　　　　图 1-21

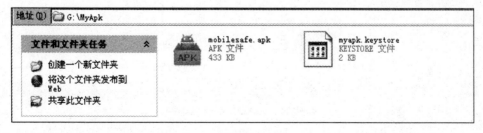

图 1-22

出现安装失败的原因在于：本地的 apk 与服务端的 apk 所采用的签名不同。所以，如果要替换安装 apk，务必要保证签名信息一致。

如果以前的程序是采用默认签名的方式（即 debug 签名），一旦换了新的签名应用将不能覆盖安装，必须将原先的程序卸载后，才能安装上。程序覆盖安装主要检查两点：

（1）两个程序的入口 Activity 是否相同。两个程序如果包名不一样，即使其他所有代码完全一样，也不会被视为同一个程序的不同版本。

（2）两个程序所采用的签名是否相同。如果两个程序所采用的签名不同，即使包名相同，也不会被视为同一个程序的不同版本，不能覆盖安装。

另外，可能有人会认为反正 debug 签名的应用程序也能安装使用，那也就没有必要自己签名了。千万不要这样想，debug 签名的应用程序有以下两个限制：

➤ debug 签名的应用程序不能在 Android Market 中上架销售，它会强制你使用自己的签名；

➤ debug.keystore 在不同的机器上所生成的可能都不一样，意味着如果你换了机器进行 apk 版本升级，那么将会出现上面那种程序不能覆盖安装的问题。不要小视这个问题，如果你开发的程序只有你自己使用，当然无所谓，卸载再安装就可以了。如果你的软件有很多用户，这就是大问题了，这就相当于软件不具备升级功能！

所以，我们不能够忽视对软件进行签名。

1.2 程序主界面的 UI 设计

当进入程序的主界面时（即 MainActivity 对应的界面），我们希望显示出如图 1-23 所示的界面效果。

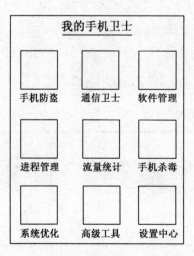

图 1-23

根据该界面效果，我们可以选择线性布局来完成其效果。由图 1-23 可知，自上而下的控件可以为：TextView（显示标题），View（显示标题下方的直线），GridView（显示手机卫士的九大模块：手机防盗、通信卫士、软件管理、进程管理、流量统计、手机杀毒、系统优化、高级工具和设置中心）。其对应的布局文件 main.xml 代码如下：

```xml
<?xml version="1.0" encoding="utf-8"?>
<LinearLayout xmlns:android="http://schemas.android.com/apk/res/android"
    android:layout_width="fill_parent"
    android:layout_height="fill_parent"
    android:orientation="vertical" >
    <TextView
        android:layout_width="fill_parent"
        android:layout_height="wrap_content"
        android:gravity="center_horizontal"
        android:text="我的手机卫士"
        android:textColor="#66ff00"
        android:textSize="28sp" />
    <View
        android:layout_width="fill_parent"
        android:layout_height="1dip"
        android:layout_marginTop="5dip"
        android:background="@drawable/devide_line" />
    <GridView
        android:id="@+id/gv_main"
        android:layout_width="fill_parent"
        android:layout_height="fill_parent"
        android:numColumns="3"
        android:layout_marginTop="5dip" >
    </GridView>
</LinearLayout>
```

代码解析：

（1）android:orientation="vertical"代表该线性布局中控件的排列方式是垂直的。

（2）android:gravity="center_horizontal"代表该属性用于设置标题"我的手机卫士"居中显示，gravity 的意思是当前的控件 TextView 中的内容相对于 TextView 本身居中显示，而 TextView 的长度设置为 android:layout_width="fill_parent"，高度设置为 android:layout_height="wrap_content"，所以内容"我的手机卫士"相对于控件本身居中显示的效果就如图 1-23 所示。

这里介绍一下 gravity 与 layout_gravity 的区别：

android:gravity 属性是对该 view 中的内容的位置进行限定（内容与当前控件的位置关系）。例如，上面的 TextView 控件，如果我们将其长度设置为 android:layout_width= "wrap_content"，其他属性保持不变，那么，标题文字将会显示在当前控件的最左端，原因在于文字本身充满了整个控件，那么也就相当于内容居中显示在控件上。

android:layout_gravity 是用来设置当前 view 相对与父 view 的位置（当前控件与父控件的位置关系）。例如，上面的 TextView 控件，如果我们将 android:gravity="center_horizontal"修改为 android:layout_gravity="center_horizontal"，保持其他属性不变，那么，标题将仍然显示在当前控件的最左端，原因在于：TextView 中的内容默认情况下是显示在控件的最左端的；TextView 的长度是 android:layout_width="fill_parent"，而 android:layout_gravity="center_horizontal"则是让当前的 TextView 控件相当于父控件<LinearLayout>居中显示，由于 TextView 的长度本身是充满父窗体的，那么，它本身也就是居中显示的。如果将 TextView 中 android:layout_width="fill_parent"修改为 android:layout_width=" wrap_content"，同时设置 android:layout_gravity="center_horizontal"，那么，标题就会居中显示了。原因在于：TextView 的长度是包裹内容的；TextView 的父控件就是<LinearLayout>（即屏幕），所以如果此时 TextView 相对于父控件居中显示，效果就是上面的居中显示样式。

对于 GridView 中的 9 个 item，我们需要为每个 item 分别添加对应的数据。每个 item 的格式都是一样的，所以我们可以先做一个 item 的基本布局。每个 item 都是由一张图片和一个标题组成的，所以对应的 main_item.xml 布局文件如下：

```xml
<?xml version="1.0" encoding="utf-8"?>
<LinearLayout xmlns:android="http://schemas.android.com/apk/res/android"
    android:layout_width="wrap_content"
    android:layout_height="wrap_content"
    android:orientation="vertical" >
    <ImageView
        android:layout_marginTop="5dip"
        android:id="@+id/iv_main_item_icon"
        android:layout_width="75dip"
        android:layout_height="75dip"
        android:src="@drawable/widget01" />
    <TextView
        android:id="@+id/tv_main_item_name"
        android:layout_width="wrap_content"
        android:layout_height="wrap_content"
        android:text="功能 1"
        android:layout_marginBottom="5dip"
        android:textSize="20sp" />
</LinearLayout>
```

此时该界面的预览界面如图 1-24 所示。

我们发现"功能 1"没能居中显示在 Image 的下面，可以在<LinearLayout>中设置一个属性 android:gravity="center_horizontal"，该属性的作用在于让其子控件都居中显示，效果图如图 1-25 所示。

图 1-24　　　　　　　　　　　图 1-25

接下来需要做的就是让每个 item 都能够显示出对应的图片与文字，此时需要借助 Android 中的 Adapter 对象来为每个 item 适配数据。MainActivity.java 中的业务代码如下：

```
package com.guoshisp.mobilesafe;
import android.app.Activity;
import android.content.Intent;
import android.os.Bundle;
import android.view.View;
import android.widget.AdapterView;
import android.widget.AdapterView.OnItemClickListener;
import android.widget.GridView;
import com.guoshisp.mobilesafe.adapter.MainAdapter;
public class MainActivity extends Activity {
    private GridView gv_main;//main.xml 中的 GridView 控件
    @Override
    protected void onCreate(Bundle savedInstanceState) {
        super.onCreate(savedInstanceState);
        setContentView(R.layout.main);
        gv_main = (GridView) findViewById(R.id.gv_main);
        //为 gv_main 对象设置一个适配器，该适配器的作用是为每个 item 填充对应的数据
        gv_main.setAdapter(new MainAdapter(this));
    }
}
```

代码解析：

gv_main.setAdapter(new MainAdapter(this))用于为 gv_main 对象设置一个适配器，该适配器要为每一个 item 填充对应的数据，在 setAdapter(ListAdapter adapter)方法中传入一个自定义的适配器对象，同时让该对象继承 BaseAdapter 对象。该对象位于 "com.guoshisp.

mobilesafe.adapter"包下。

适配器对象 MainAdapter.java 中的代码如下:

```java
package com.guoshisp.mobilesafe.adapter;
import android.content.Context;
import android.view.LayoutInflater;
import android.view.View;
import android.view.ViewGroup;
import android.widget.BaseAdapter;
import android.widget.ImageView;
import android.widget.TextView;
import com.guoshisp.mobilesafe.R;
public class MainAdapter extends BaseAdapter {
    //布局填充器
    private LayoutInflater inflater;
    //接收 MainActivity 传递过来的上、下文对象
    private Context context;
    //将9个 item 的每一个图片对应的 id 都存入该数组中
    private static final int[] icons = { R.drawable.widget01,
        R.drawable.widget02, R.drawable.widget03, R.drawable.widget04,
        R.drawable.widget05, R.drawable.widget06, R.drawable.widget07,
        R.drawable.widget08, R.drawable.widget09 };
    //将9个 item 的每一个标题都存入该数组中
    private static final String[] names = { "手机防盗", "通信卫士", "软件管理", "进程管理","流量统计", "手机杀毒", "系统优化", "高级工具", "设置中心"};
    public MainAdapter(Context context) {
        this.context = context;
        //获取系统中的布局填充器
        inflater = (LayoutInflater) context.getSystemService(Context.LAYOUT_INFLATER_SERVICE);
    }
    /**
     * 返回 gridview 有多少个 item
     */
    public int getCount() {
        return names.length;
    }
    /**
     * 获取每个 item 对象,如果不对这个返回的 item 对象做相应的操作
     * 可以返回一个 null,这里我们简单处理一下,返回 position
     */
    public Object getItem(int position) {
        return position;
    }
    /**
```

```
     * 返回当前 item 的 id
     */
    public long getItemId(int position) {
        return position;
    }
    /**
     * 返回每一个 gridview 条目中的 view 对象
     */
    public View getView(int position, View convertView, ViewGroup parent) {
        View view = inflater.inflate(R.layout.main_item, null);
        TextView tv_name = (TextView) view.findViewById(R.id.tv_main_
                            item_name);
        ImageView iv_icon = (ImageView) view.findViewById(R.id.iv_main_
                            item_icon);
        tv_name.setText(names[position]);
        iv_icon.setImageResource(icons[position]);
        return view;
    }
}
```

代码解析：

（1）inflater = (LayoutInflater) context
　　　　.getSystemService(Context.LAYOUT_INFLATER_SERVICE)

获取系统的布局填充器，布局填充器可以将一个布局文件转换成 View 对象。

（2）getView(int position, View convertView, ViewGroup parent)

该方法是用于为每个 item 返回一个对应的 View，该方法中的 position 就是当前 item 对应的 id。

（3）View view = inflater.inflate(R.layout.main_item, null)

通过布局填充器将布局文件转成 View 对象，该 View 对象也是我们需要为对应的 item 返回的对象。这里我们用不到 ViewGroup 对象，所以将其设置为 null。TextView tv_name = (TextView)view.findViewById(R.id.tv_main_item_name)与 ImageView iv_icon = (ImageView) view.findViewById(R.id.iv_main_item_icon)获取到 main_item.xml 中的 TextView 对象。需要特别注意的一点是：在调用 findViewById(int id)时，一定要指定是获取哪个 View 中的控件 id，否则会出现找不到该 id 的错误，因为默认是在当前的 Activity 中查找控件的 id 的。

（4）tv_name.setText(names[position])
　　　iv_icon.setImageResource(icons[position])

分别为 TextView 和 ImageView 设置对应的文字与图片。

（5）return view 将填充好数据的 item 对应的 View 对象返回给对应的 item。

运行程序，主界面的效果图如图 1-26 所示。

但是，当我们单击其中的任意一个图标时，图标的背景色没有发生改变，只是字体颜色变淡一点，效果图如图 1-27 所示。如果将其运行在 Android2.2 的环境下，当我们单击每一个图标时，背景色是橘黄色的，与我们的 UI 不搭配。所以，为了让 UI 看起来更加美观，我们需要重写系统默认选择后的背景色，然后需要将系统默认的颜色禁用。

因为每个 item 对应的 View 是通过 main_item.xml 布局文件转化过来的，所以，为了解决这个问题，我们需要修改 main_item.xml 布局文件，修改后的布局文件如下：

```xml
<?xml version="1.0" encoding="utf-8"?>
<LinearLayout xmlns:android="http://schemas.android.com/apk/res/android"
    android:layout_width="wrap_content"
    android:layout_height="wrap_content"
    android:gravity="center_horizontal"
    android:background="@drawable/bg_selector"
    android:orientation="vertical" >
    <ImageView
        android:layout_marginTop="5dip"
        android:id="@+id/iv_main_item_icon"
        android:layout_width="75dip"
        android:layout_height="75dip"
        android:src="@drawable/widget01" />
    <TextView
        android:id="@+id/tv_main_item_name"
        android:layout_width="wrap_content"
        android:layout_height="wrap_content"
        android:text="功能 1"
        android:layout_marginBottom="5dip"
        android:textSize="20sp" />
</LinearLayout>
```

可以看到，我们在<LinearLayout>中添加了 android:background="@drawable/ bg_selector"，可以通过这个设置来控制 View 被单击、获取焦点、默认时的各种状态。

进入 SDK 目录下的 docs 目录,使用浏览器打开 docs 目录下的 index.xml 文件后,单击"Dev Guide"，进入 Application Resource\Resource Types\Drawable 目录，然后在右边显示的内容中找到"State List"（状态列表），通过查看 API 我们可以定义 View 被按下、获取焦点、正常状态时的不同显示状态。我们将定义好的 XML 文件存放在 drawable 目录下。

bg_selector.xml 对应的代码为：

```xml
<?xml version="1.0" encoding="utf-8"?>
<selector xmlns:android="http://schemas.android.com/apk/res/android">
    <item android:state_pressed="true"
        android:drawable="@drawable/bg_normal_pressed" /> <!-- pressed -->
    <item android:state_focused="true"
```

```xml
        android:drawable="@drawable/bg_normal_pressed" /> <!-- focused -->
    <item android:drawable="@drawable/bg_normal" /> <!-- default -->
</selector>
```
其中的 bt_selected.xml 和 bt_normal.xml 都存放在 drawable 目录下

bg_normal_pressed.xml 对应的代码为：

```xml
<?xml version="1.0" encoding="utf-8"?>
<shape xmlns:android="http://schemas.android.com/apk/res/android"
    android:shape="rectangle" >
    <corners android:radius="4dip" >
    </corners>
    <solid android:color="#5b5b5b" />
</shape>
```

代码解析：

（1）android:shape="rectangle"用于指定单击 View 时所显示的背景图形，这里指定的是矩形。

（2）<corners android:radius="4dip" >

　　</corners>

用于指定矩形的四个角的角度。

（3）<solid android:color="#5b5b5b" />

　　</shape>

用于指定矩形的颜色。

bg_normal.xml 对应的代码为：

```xml
<?xml version="1.0" encoding="utf-8"?>
<shape xmlns:android="http://schemas.android.com/apk/res/android"
    android:shape="rectangle" >
    <corners android:radius="4dip" >
    </corners>
    <solid android:color="#000000" />
</shape>
```

禁用系统默认的背景色，可以修改 main.xml 中的 GridView 控件，在该控件的第一行添加一个属性，该属性是将系统默认的橘黄色改为透明色，代码如下：

```xml
<GridView
    android:listSelector="@android:color/transparent"
    android:id="@+id/gv_main"
    android:layout_width="fill_parent"
    android:layout_height="fill_parent"
    android:numColumns="3"
    android:layout_marginTop="5dip" >
</GridView>
```

修改后，运行并单击 item 时对应的效果图如图 1-28 所示。

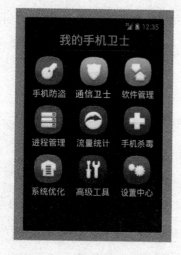
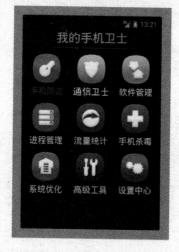
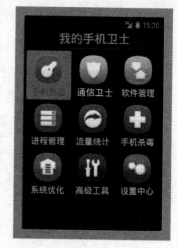

图 1-26　　　　　　　　　图 1-27　　　　　　　　　图 1-28

1.3　关闭自动更新

如果服务端有最新版本时,当我们完全退出程序后再次进入时,总会弹出升级对话框,或者我们不想让手机执行连网更新动作,可以在设置中心模块中关闭软件的自动更新功能。我们希望"设置中心"界面效果如图 1-29 所示。

所以界面的主要控件有 TextView、两个 View(分别显示实线和虚线)和 Checkbox。当 Checkbox 处于勾选状态时,对应的"自动更新没有开启"应该修改为"自动更新已开启",并将此时的字体颜色从红色修改为白色。默认情况下,自动更新时是开启的。界面对应的布局文件为 setting_center.xml,其对应的代码如下:

```xml
<?xml version="1.0" encoding="utf-8"?>
<LinearLayout xmlns:android="http://schemas.android.com/apk/res/android"
    android:layout_width="fill_parent"
    android:layout_height="fill_parent"
    android:orientation="vertical" >
<TextView
    android:layout_width="wrap_content"
    android:layout_height="wrap_content"
    android:layout_gravity="center_horizontal"
    android:text="设置中心"
    android:textColor="#66ff00"
    android:textSize="28sp" />
<View
    android:layout_width="fill_parent"
    android:layout_height="1dip"
    android:layout_marginTop="5dip"
    android:background="@drawable/devide_line" />
<RelativeLayout
```

```
            android:layout_marginTop="8dip"
            android:layout_width="fill_parent"
            android:layout_height="wrap_content" >
            <TextView
                android:id="@+id/tv_setting_autoupdate_text"
                android:layout_width="wrap_content"
                android:layout_height="wrap_content"
                android:text="自动更新设置"
                android:textSize="26sp"
                android:textStyle="bold" >
            </TextView>
            <TextView
                android:id="@+id/tv_setting_autoupdate_status"
                android:layout_width="wrap_content"
                android:layout_height="wrap_content"
                android:layout_below="@id/tv_setting_autoupdate_text"
                android:text="自动更新没有开启"
                android:textColor="#ffffff"
                android:textSize="14sp" />
            <CheckBox
                android:id="@+id/cb_setting_autoupdate"
                android:layout_width="wrap_content"
                android:layout_height="wrap_content"
                android:layout_alignParentRight="true"
                android:checked="false" />
        </RelativeLayout>
        <View
            android:layout_marginTop="4dip"
            android:layout_width="fill_parent"
            android:layout_height="1dip"
            android:background="@drawable/listview_devider" >
        </View>
    </LinearLayout>
```

对应的预览图如图 1-30 所示。

图 1-29

图 1-30

其对应的 Activity 为 SettingCenterActivity，然后在清单文件中配置对应的信息，SettingCenterActivity.java 对应的业务代码如下：

```java
package com.guoshisp.mobilesafe;
import android.app.Activity;
import android.content.SharedPreferences;
import android.content.SharedPreferences.Editor;
import android.graphics.Color;
import android.os.Bundle;
import android.widget.CheckBox;
import android.widget.CompoundButton;
import android.widget.CompoundButton.OnCheckedChangeListener;
import android.widget.TextView;
public class SettingCenterActivity extends Activity {
    //用于存储自动更新是否开启的boolean值
    private SharedPreferences sp;
    //自动更新是否开启对应的TextView控件的显示文字
    private TextView tv_setting_autoupdate_status;
    //显示自动更新是否开启的勾选框
    private CheckBox cb_setting_autoupdate;
    @Override
    protected void onCreate(Bundle savedInstanceState) {
        setContentView(R.layout.setting_center);
        super.onCreate(savedInstanceState);
        //获取Sdcard下的config.xml文件，如果该文件不存在，那么将会自动创建该文
        //  件，文件的操作类型为私有类型
        sp = getSharedPreferences("config", MODE_PRIVATE);
        //标记自动更新状态是否开启对应的Checkbox控件
        cb_setting_autoupdate = (CheckBox)findViewById(R.id.cb_setting_
                        autoupdate);
        //显示当前自动更新是否开启对应的TextView控件
        tv_setting_autoupdate_status = (TextView)findViewById(R.id.tv_
                        setting_autoupdate_status);
        //初始化自动更新的UI，默认状态下是开启的
        boolean autoupdate = sp.getBoolean("autoupdate", true);
        if(autoupdate){
            tv_setting_autoupdate_status.setText("自动更新已经开启");
            //因为autoupdate变量为true，则表示自动更新开启，所以，Checkbox的
            //  状态应该是勾选状态的，即为true
            cb_setting_autoupdate.setChecked(true);
        }else{
            tv_setting_autoupdate_status.setText("自动更新已经关闭");
            //因为autoupdate变量为false，则表示自动更新未开启，所以，Checkbox
            //  的状态应该是未勾选状态的，即为false
            cb_setting_autoupdate.setChecked(false);
        }
        /**
```

```
 * 当Checkbox的状态发生改变时执行以下代码
 */
cb_setting_autoupdate.setOnCheckedChangeListener(new
    OnCheckedChangeListener() {
        //参数一：当前的Checkbox；第二个参数：当前的Checkbox是否处于勾选状态
        public void onCheckedChanged(CompoundButton buttonView, boolean
                                    isChecked) {
            //获取编辑器
            Editor editor = sp.edit();
            //持久化存储当前Checkbox的状态，当下次进入时，依然可以保存当前设置的状态
            editor.putBoolean("autoupdate", isChecked);
            //将数据真正提交到sp里面
            editor.commit();
            if(isChecked){//Checkbox处于选中效果
            //当Checkbox处于勾选状态时，表示自动更新已经开启，同时修改字体颜色
                tv_setting_autoupdate_status.setTextColor(Color.WHITE);
                tv_setting_autoupdate_status.setText("自动更新已经开启");
            }else{//Checkbox处于未勾选状态
            //当Checkbox未处于勾选状态时，表示自动更新已经开启，同时修改字体颜色
                tv_setting_autoupdate_status.setTextColor(Color.RED);
                tv_setting_autoupdate_status.setText("自动更新已经关闭");
            }
        }
    });
}
}
```

代码解析：

（1）sp = getSharedPreferences("config", MODE_PRIVATE)

在手机文件的 data\data\应用程序包名\share_prefs 目录下创建一个 config.xml 文件，文件的操作类型为 MODE_PRIVATE。

（2）boolean autoupdate = sp.getBoolean("autoupdate", true)

从 sp 对应的 config.xml 文件中获取键 autoupdate 所对应的 boolean 值，如果没有查找到该键，将会返回一个默认的 boolean 值——true。

（3）Editor editor = sp.edit()

获取到一个编辑器对象 Editor，该对象可以向 sp 中存入数据。

（4）editor.putBoolean("autoupdate", isChecked)

用 sp 所对应的编辑器对象 editor 向 config.xml 中存入数据，数据是以键值对的方式存放的，前面是键，后面是键所对应的值，当需要取出时，可以通过键来获取对应的值。

（5）editor.commit()

将编辑器编辑的数据真正地提交到 sp 对应的 config.xml 文件中。如果没有这句代码，数据

将不会被保存到 config.xml 文件中，也就不能够取出对应的数据。

"设置中心"界面及逻辑已经处理完毕，接下来，当我们在主界面中单击"设置中心"模块时应该进入该界面，所以在 MainActivity 中需要为其设置单击的监听事件。此时 MainActivity.java 对应的业务代码如下：

```java
package com.guoshisp.mobilesafe;
import android.app.Activity;
import android.content.Intent;
import android.os.Bundle;
import android.view.View;
import android.widget.AdapterView;
import android.widget.AdapterView.OnItemClickListener;
import android.widget.GridView;
import com.guoshisp.mobilesafe.adapter.MainAdapter;
public class MainActivity extends Activity {
    //显示主界面中的九大模块的GridView
    private GridView gv_main;
    @Override
    protected void onCreate(Bundle savedInstanceState) {
        super.onCreate(savedInstanceState);
        setContentView(R.layout.main);
        gv_main = (GridView) findViewById(R.id.gv_main);
        //为gv_main对象设置一个适配器，该适配器的作用是为每个item填充对应的数据
        gv_main.setAdapter(new MainAdapter(this));
        //为GridView对象中的item设置单击时的监听事件
        gv_main.setOnItemClickListener(new OnItemClickListener() {
            //参数一：item的父控件，也就是GridView；
            //参数二：当前单击的item
            //参数三：当前单击的item在GridView中的位置
//参数四：id的值为单击GridView的哪一项对应的数值，单击GridView第9项，那id
    就等于8
            public void onItemClick(AdapterView<?> parent, View view,
                    int position, long id) {
                switch (position) {
                case 8://设置中心
                    //跳转到"设置中心"对应的Activity界面
                    Intent settingIntent = new Intent(MainActivity.this,
                            SettingCenterActivity.class);
                    startActivity(settingIntent);
                    break;
                }
            }
        });
    }
}
```

此时，运行程序，单击"设置中心"模块后的显示界面如图 1-31 所示，接着进入 data\data\com.

guoshisp.mobilesafe\share_prefs 目录下导出并打开 config.xml 文件，该文件对应的内容如下：

```xml
<?xml version='1.0' encoding='utf-8' standalone='yes' ?>
<map>
<boolean name="autoupdate" value="true" />
</map>
```

图 1-31

所以，此时可以在 SplashActivity 中根据"设置中心"中 Checkbox 的状态来判断是否需要连网更新。

SplashActivity.java 对应的业务代码见在线资源包中代码文本文件 1.3.doc。

当我们进入"设置中心"取消自动更新设置，完全退出程序（通过 back 键退出），并修改服务端的版本号后，此时，再次进入应用程序时，就不会出现升级提示对话框了。说明"设置中心"中的设置起作用了。

第 2 章 手机防盗模块的设计

2.1 手机防盗的功能介绍

功能介绍：当首次进入"手机防盗"界面时，需要为手机设置防盗密码，设置完成后，当再次进入时需要输入手机防盗的密码。当正确输入密码后，就进入了手机防盗的设置向导界面（与其他类似软件的设置向导相似，如 QQ 输入法的设置向导），接下来就是要根据设置向导来完成手机防盗安全号码的绑定和激活手机防盗功能。安全号码一旦绑定，并且激活了手机防盗，那么我们就可以通过绑定的安全号码向手机发送一些防盗指令——获取手机位置、播放报警音乐、远程锁屏、清除数据。在以后每次成功进入手机防盗界面后，界面会显示出安全号码、防盗保护设置是否开启、重新进入设置向导、手机防盗指令。

UI 设置：当第一次进入"手机防盗"时，在屏幕中央弹出一个对话框，如图 2-1 所示。设置好密码后，单击"确定"按钮后立即进入主界面，当再次进入"手机防盗"时，在屏幕中央会弹出一个对话框，如图 2-2 所示。正确输入密码后会进入设置向导界面，根据设置向导来完成"手机防盗"的具体设置。完成设置向导中的设置后，当以后再次成功进入"手机防盗"界面时，会显示设置信息。

图 2-1

图 2-2

"手机防盗"对应的 LostProtectedActivity 的业务代码如下（同时在清单文件中填加相对应

的配置信息）：

```java
package com.guoshisp.mobilesafe;
import android.app.Activity;
import android.app.AlertDialog;
import android.app.AlertDialog.Builder;
import android.content.DialogInterface;
import android.content.DialogInterface.OnCancelListener;
import android.content.SharedPreferences;
import android.content.SharedPreferences.Editor;
import android.os.Bundle;
import android.text.TextUtils;
import android.view.Menu;
import android.view.View;
import android.view.View.OnClickListener;
import android.widget.Button;
import android.widget.EditText;
import android.widget.Toast;
import com.guoshisp.mobilesafe.utils.Md5Encoder;
public class LostProtectedActivity extends Activity implements
        OnClickListener {
    //偏好设置存储对象
    private SharedPreferences sp;
    //第一次进入"手机防盗"界面时的界面控件对象
    private EditText et_first_dialog_pwd;
    private EditText et_first_dialog_pwd_confirm;
    private Button bt_first_dialog_ok;
    private Button bt_first_dialog_cancle;
    //第二次进入"手机防盗"界面时的界面控件对象
    private EditText et_normal_dialog_pwd;
    private Button bt_normal_dialog_ok;
    private Button bt_normal_dialog_cancle;
    //对话框对象
    private AlertDialog dialog;
    @Override
    protected void onCreate(Bundle savedInstanceState) {
        super.onCreate(savedInstanceState);
        //获取Sdcard下的config.xml文件,如果该文件不存在,那么将会自动创建该文件
        sp = getSharedPreferences("config", MODE_PRIVATE);
        //判断用户是否设置过密码
        if (isSetupPwd()) {
            //进入非第一次进入"手机防盗"时要显示的对话框
            showNormalEntryDialog();
        } else {
            //进入第一次进入"手机防盗"时要显示的对话框
            showFirstEntryDialog();
```

```java
    }
}
/**
 * 第一次进入"手机防盗"时要显示的对话框
 */
private void showFirstEntryDialog() {
    //得到对话框的构造器
    AlertDialog.Builder builder = new Builder(this);
    //通过View对象的inflate(Context context, int resource, ViewGroup
      root)对象将第一次进入"手机防盗"要弹出的窗体对话框的布局文件转换为一个
      View对象
    View view = View.inflate(this, R.layout.first_entry_dialog, null);
    //查找view对象中的各个控件
    et_first_dialog_pwd = (EditText) view
            .findViewById(R.id.et_first_dialog_pwd);
    et_first_dialog_pwd_confirm = (EditText) view
            .findViewById(R.id.et_first_dialog_pwd_confirm);
    bt_first_dialog_ok = (Button) view
            .findViewById(R.id.bt_first_dialog_ok);
    bt_first_dialog_cancle = (Button) view
            .findViewById(R.id.bt_first_dialog_cancle);
    //分别为"取消"、"确定"按钮设置一个监听器
    bt_first_dialog_cancle.setOnClickListener(this);
    bt_first_dialog_ok.setOnClickListener(this);
    //将上面的View对象添加到对话框上
    builder.setView(view);
    //获取到对话框对象
    dialog = builder.create();
    //显示出对话框
    dialog.show();
}
/**
 * 当设置过密码后,正常进入"手机防盗"时要显示的对话框
 */
private void showNormalEntryDialog() {
    AlertDialog.Builder builder = new Builder(this);
    builder.setOnCancelListener(new OnCancelListener() {
//当单击"取消"按钮时,直接结束当前的LostProtectedActivity,程序会进入到主界面
        public void onCancel(DialogInterface dialog) {
            finish();
        }
    });
    //通过View对象的inflate(Context context, int resource, ViewGroup
      root)对象将非第一次进入"手机防盗"要弹出的窗体对话框的布局文件转换为一个
      View对象
    View view = View.inflate(this, R.layout.normal_entry_dialog, null);
```

```java
        //查找view对象中的各个控件
        et_normal_dialog_pwd = (EditText) view
                .findViewById(R.id.et_normal_dialog_pwd);
        bt_normal_dialog_ok = (Button) view
                .findViewById(R.id.bt_normal_dialog_ok);
        bt_normal_dialog_cancle = (Button) view
                .findViewById(R.id.bt_normal_dialog_cancle);
        //分别为"取消"、"确定"按钮设置一个监听器
        bt_normal_dialog_cancle.setOnClickListener(this);
        bt_normal_dialog_ok.setOnClickListener(this);
        //将上面的View对象添加到对话框上
        builder.setView(view);
        //获取到对话框对象
        dialog = builder.create();
        //显示出对话框
        dialog.show();
    }
    /**
     * 判断用户是否设置过密码
     *
     * @return
     */
    private boolean isSetupPwd() {
        String savedpwd = sp.getString("password", "");
        if (TextUtils.isEmpty(savedpwd)) {//通过一个文本工具类来判断String是否为空
            return false;
        } else {
            return true;
        }
        // return (!TextUtils.isEmpty(savedpwd));
    }
    /**
     * 为两个对话框中的"确定"和"取消"按钮设置的监听器
     */
    public void onClick(View v) {
        switch (v.getId()) {
        //第一次进入"手机防盗"时弹出的对话框中，对"取消"按钮事件的处理
        case R.id.bt_first_dialog_cancle:
            dialog.cancel();//取消对话框
            finish();//结束当前的Activity后会进入程序的主界面
            break;
        //第一次进入"手机防盗"时弹出的对话框中，对"确定"按钮事件的处理
        case R.id.bt_first_dialog_ok:
            //获取到两个EditText中的输入的密码，并将EditText前后的空格去除
            String pwd = et_first_dialog_pwd.getText().toString().trim();
            String pwd_confirm = et_first_dialog_pwd_confirm.getText()
```

```java
            .toString().trim();
    //判断两个EditText中的内容是否为空
    if (TextUtils.isEmpty(pwd_confirm) || TextUtils.isEmpty(pwd)) {
        Toast.makeText(this, "密码不能为空", 1).show();
        return;
    }
    //判断两个EditText中的内容是否相同
    if (pwd.equals(pwd_confirm)) {
        //获取到一个编辑器对象,此处用于向sp中编辑数据
        Editor editor = sp.edit();
        //将加密后的密码存入到sp所对应的文件中
        editor.putString("password", Md5Encoder.encode(pwd));
        //将编辑的数据提交后才能真正地存入sp中
        editor.commit();
        //销毁当前的对话框
        dialog.dismiss();
        //结束当前的Activity后,跳转至主界面
        finish();
    } else {
        Toast.makeText(this, "两次密码不相同", 1).show();
        return;
    }
    break;
    //非第一次进入"手机防盗"时弹出的对话框中,对"取消"按钮事件的处理
case R.id.bt_normal_dialog_cancle:
    dialog.cancel();
    finish();
    break;
    //非第一次进入"手机防盗"时弹出的对话框中,对"确定"按钮事件的处理
case R.id.bt_normal_dialog_ok:
    String userentrypwd = et_normal_dialog_pwd.getText().toString()
                        .trim();
    if (TextUtils.isEmpty(userentrypwd)) {
        Toast.makeText(this, "密码不能为空", 1).show();
        return;
    }
    String savedpwd = sp.getString("password", "");
    //因为在设置密码后,存入的是加密后的密码,所以当我们将输入的密码与设
    //  置的密码比较时需要将输入的密码先加密
    if (savedpwd.equals(Md5Encoder.encode(userentrypwd))) {
        Toast.makeText(this, "密码正确进入界面", 1).show();
        dialog.dismiss();
        //加载主界面
        return;
    } else {
        Toast.makeText(this, "密码不正确", 1).show();
```

```
            return;
         }
      }
   }
}
```

代码解析：

（1）String savedpwd = sp.getString("password", "")

用来获取 sp 中存放的 String 类型的密码，也就是说，在我们存储密码时也是以 String 类型来存入的。之所以以 String 来存储是因为在 Android 中，手工输入的所有信息都是 String 类型的。而且，默认的是空，这也正好为我们后面使用 TextUtils.isEmpty (savedpwd)提供了方便。

（2）editor.putString("password", Md5Encoder.encode(pwd))

当设置完防盗密码后，存储时需要将密码加密后再存储到 sp 对应的文件中。这里使用的是 MD5 加密算法对密码进行加密。

接下来介绍 LostProtectedActivity.java 中所使用的几个布局文件（全部存放在 layout 文件夹下）和工具类代码。

first_entry_dialog.xml 布局文件代码如下：

```xml
<?xml version="1.0" encoding="utf-8"?>
<LinearLayout xmlns:android="http://schemas.android.com/apk/res/android"
    android:layout_width="260dip"
    android:layout_height="wrap_content"
    android:gravity="center_horizontal"
    android:orientation="vertical" >
    <TextView
        android:layout_width="wrap_content"
        android:layout_height="wrap_content"
        android:text="设置密码"
        android:textColor="#66ff00"
        android:textSize="26sp" />
    <View
        android:layout_width="fill_parent"
        android:layout_height="1dip"
        android:layout_marginTop="5dip"
        android:background="@drawable/devide_line" >
    </View>
    <EditText
        android:id="@+id/et_first_dialog_pwd"
        android:layout_width="fill_parent"
        android:inputType="textPassword"
        android:layout_height="wrap_content"
        android:hint="请输入密码" />
    <EditText
```

```xml
        android:id="@+id/et_first_dialog_pwd_confirm"
        android:layout_width="fill_parent"
        android:inputType="textPassword"
        android:layout_height="wrap_content"
        android:hint="请再次输入密码" />
    <LinearLayout
        android:layout_width="260dip"
        android:layout_height="wrap_content"
        android:orientation="horizontal" >
        <Button
            android:layout_width="0dip"
            android:layout_weight="1"
            android:layout_height="wrap_content"
            android:id="@+id/bt_first_dialog_ok"
            android:text="确定" />
        <Button
            android:id="@+id/bt_first_dialog_cancle"
            android:layout_width="0dip"
            android:layout_weight="1"
            android:layout_height="wrap_content"
            android:text="取消" />
    </LinearLayout>
</LinearLayout>
```

normal_entry_dialog.xml 布局文件代码如下：

```xml
<?xml version="1.0" encoding="utf-8"?>
<LinearLayout xmlns:android="http://schemas.android.com/apk/res/android"
    android:layout_width="260dip"
    android:layout_height="wrap_content"
    android:gravity="center_horizontal"
    android:orientation="vertical" >
    <TextView
        android:layout_width="wrap_content"
        android:layout_height="wrap_content"
        android:text="输入密码"
        android:textColor="#66ff00"
        android:textSize="26sp" />
    <View
        android:layout_width="fill_parent"
        android:layout_height="1dip"
        android:layout_marginTop="5dip"
        android:background="@drawable/devide_line" >
    </View>
    <EditText
        android:id="@+id/et_normal_dialog_pwd"
        android:layout_width="fill_parent"
```

```xml
        android:layout_height="wrap_content"
        android:inputType="textPassword"
        android:hint="请输入密码" />
    <LinearLayout
        android:layout_width="260dip"
        android:layout_height="wrap_content"
        android:orientation="horizontal" >
        <Button
            android:layout_width="0dip"
            android:layout_weight="1"
            android:layout_height="wrap_content"
            android:id="@+id/bt_normal_dialog_ok"
            android:text="确定" />
        <Button
            android:id="@+id/bt_normal_dialog_cancle"
            android:layout_width="0dip"
            android:layout_weight="1"
            android:layout_height="wrap_content"
            android:text="取消" />
    </LinearLayout>
</LinearLayout>
```

工具类代码 **Md5Encoder.java** 存放在包 "**com.guoshisp.mobilesafe.utils**" 下，代码如下：

```java
package com.guoshisp.mobilesafe.utils;
import java.security.MessageDigest;
import java.security.NoSuchAlgorithmException;
public class Md5Encoder {
    public static String encode(String password){
        try {
            //获取到数字消息的摘要器
            MessageDigest digest = MessageDigest.getInstance("MD5");
            //执行加密操作
            byte[] result = digest.digest(password.getBytes());
            StringBuilder sb = new StringBuilder();
            //将每个byte字节的数据转换成十六进制的数据
            for(int i= 0 ;i<result.length;i++){
                int number = result[i]&0xff;
                String str = Integer.toHexString(number);//将十进制的number
                            转换成十六进制数据
                if(str.length()==1){//判断加密后的字符的长度，如果长度为1，则
                            在该字符前面补0
                    sb.append("0");
                    sb.append(str);
                }else{
                    sb.append(str);
                }
```

```
            }
            return sb.toString();//将加密后的字符转成字符串返回
        } catch (NoSuchAlgorithmException e) {//加密器没有被找到,该异常不可
          能发生,因为填入的"MD5"是正确的
            e.printStackTrace();
            return "";
        }
    }
}
```

当手机丢失后,如果是"普通人"拾到手机,他可能没有办法进入我们的手机防盗界面,但如果对方是专业人士,且手机密码没有被加密,他很轻松的就可以获取到我们设置的密码,然后进入手机防盗界面去修改我们绑定的安全号码。这样,"手机防盗"功能就失去了作用。

MD5 加密有两个特点:

➢ 输入两个不同的明文不会得到相同的输出值;

➢ 根据输出值不能得到原始的明文,即"不可逆"。

所以要破解 MD5 加密,还没有现成的算法,只能用穷举法,把可能出现的明文,用 MD5 算法散列之后,将得到的散列值和原始的数据形成一个一对一的映射表,所谓的解密,都是通过这个映射表来查找其所对应的原始明文。目笔者了解的情况而言,还没有一种算法可以通过输出加密后的散列值算出原始明文。

完成 LostProtectedActivity 中对应的逻辑后,我们还需要在 MainActivity 中为"手机防盗"设置单击事件的监听处理,代码如下:

```
package com.guoshisp.mobilesafe;
import android.app.Activity;
import android.content.Intent;
import android.os.Bundle;
import android.view.View;
import android.widget.AdapterView;
import android.widget.AdapterView.OnItemClickListener;
import android.widget.GridView;
import com.guoshisp.mobilesafe.adapter.MainAdapter;
public class MainActivity extends Activity {
    //显示主界面中九大模块的GridView
    private GridView gv_main;
    @Override
    protected void onCreate(Bundle savedInstanceState) {
        super.onCreate(savedInstanceState);
        setContentView(R.layout.main);
        gv_main = (GridView) findViewById(R.id.gv_main);
        //为gv_main对象设置一个适配器,该适配器的作用是为每个item填充对应的数据
        gv_main.setAdapter(new MainAdapter(this));
        //为GridView对象中的item设置单击时的监听事件
```

```
gv_main.setOnItemClickListener(new OnItemClickListener() {
    //参数一: item 的父控件, 也就是 GridView; 参数二: 当前单击的 ite; 参数
        三: 当前单击的 item 在 GridView 中的位置
    //参数四: id 的值为单击 GridView 的哪一项对应的数值, 单击 GridView
        第9项, 那 id 就等于 8
    public void onItemClick(AdapterView<?> parent, View view,
        int position, long id) {
        switch (position) {
        case 0: //手机防盗
            //跳转到"手机防盗"对应的 Activity 界面
            Intent lostprotectedIntent = new Intent(MainActivity.
                                this,LostProtectedActivity.class);
            startActivity(lostprotectedIntent);
            break;
        case 8://设置中心
            //跳转到"设置中心"对应的 Activity 界面
            Intent settingIntent = new Intent(MainActivity.this,
                                SettingCenterActivity.class);
            startActivity(settingIntent);
            break;
        }
    }
});
}
```

测试运行, 进入主界面后单击"手机防盗", 出现界面第一次需要弹出的对话框, 设置密码为"123456", 如图 2-1 所示。当再次进入"手机防盗"时, 弹出的对话框如图 2-2 所示。

2.2 手机防盗的细节

假如我们的手机被小偷偷走了, 当他打开手机进入手机安全卫士时看到有"手机防盗"这么一个模块, 那么他肯定会卸载掉手机安全卫士。为了降低该软件被卸载掉的可能性, 这里列出两种方案。

方案一: 最好能够修改"手机防盗"的字样来迷惑一下对方。例如, 当我们成功进入"手机防盗"的主界面时, 长按 Menu 键会弹出一个"修改标题"的提示, 可将"手机防盗"修改为"MP3"或其他标题。修改之后, 当下次再次进入程序时, 修改就会生效。

方案二: 采用拨号的形式来进入"手机防盗"界面, 而在程序的主界面中隐藏掉该功能(为了保持主界面的九宫格的风格, 只实现拨号进入的方式, 读者如果有兴趣的话, 可以自己动手实现一下)。

方案一的实现: 既然是需要在成功进入"手机防盗"模块后才可以进行修改标题, 那么要

实现该功能就需要进入到 LostProtectedActivity 中来实现，这里需要实现 onCreateOptionsMenu (Menu menu)和 onOptionsItemSelected(MenuItem item)方法，实现后的 LostProtectedActivity.java 的业务代码见在线资源包中代码文本部分 2.2.doc。

说明：在 onOptionsItemSelected(MenuItem item)方法中可以看到，当单击窗体对话框的"确定"按钮时，我们将文本输入框中的内容保存到了 sp 中，而此时"手机防盗"标题并没有发生改变，因为为"手机防盗"这个 Item 设置标题时是在程序加载主界面（MainActivity）时执行的，具体是在加载主界面时为主界面中对应的 GridView 中的每个 Item 设置对应的数据时实现的。所以应该在 MainActivity 中的 GridView 对应的适配器 MainAdapter.java 中来实现标题的修改，具体代码如下：

```java
package com.guoshisp.mobilesafe.adapter;
import android.content.Context;
import android.content.SharedPreferences;
import android.text.TextUtils;
import android.view.LayoutInflater;
import android.view.View;
import android.view.ViewGroup;
import android.widget.BaseAdapter;
import android.widget.ImageView;
import android.widget.TextView;
import com.guoshisp.mobilesafe.R;
public class MainAdapter extends BaseAdapter {
    //布局填充器
    private LayoutInflater inflater;
    //接收 MainActivity 传递过来的上、下文对象
    private Context context;
    //用于替换"手机防盗"的新标题
    private String newname;
    private static final int[] icons = { R.drawable.widget01,
            R.drawable.widget02, R.drawable.widget03, R.drawable.widget04,
            R.drawable.widget05, R.drawable.widget06, R.drawable.widget07,
            R.drawable.widget08, R.drawable.widget09 };
    //将9个item的每一个标题都存入该数组中
    private static final String[] names = { "手机防盗", "通信卫士", "软件管
            理", "进程管理","流量统计", "手机杀毒", "系统优化", "高级工具", "设置中心" };
    public MainAdapter(Context context) {
        this.context = context;
        //获取系统中的布局填充器
        inflater = (LayoutInflater) context
                .getSystemService(Context.LAYOUT_INFLATER_SERVICE);
        //获取用于替换"手机防盗"的新标题，默认值为空
        SharedPreferences sp = context.getSharedPreferences("config",
                Context.MODE_PRIVATE);
```

```java
        newname = sp.getString("newname", "");
    }
    /**
     * 返回gridview有多少个item
     */
    public int getCount() {
        return names.length;
    }
    /**
     * 获取每个item对象，如果不对这个返回的item对象做相应的操作，可以返回
     *   一个null，这里我们简单处理一下，返回position
     */
    public Object getItem(int position) {
        return position;
    }
    /**
     * 返回当前item的id
     */
    public long getItemId(int position) {
        return position;
    }
    /**
     * 返回每一个gridview中条目中的view对象
     */
    public View getView(int position, View convertView, ViewGroup parent) {
        View view = inflater.inflate(R.layout.main_item, null);
        TextView tv_name = (TextView) view.findViewById(R.id.tv_main_item_
                        name);
        ImageView iv_icon = (ImageView) view
                .findViewById(R.id.iv_main_item_icon);
        tv_name.setText(names[position]);
        iv_icon.setImageResource(icons[position]);
        //第一个Item，即"手机防盗"对应的Item
        if (position == 0) {
            //判断sp中取出的newname是否为空，如果不为空，将"手机防盗"对应的标
                题修改为sp中修改后的标题
            if (!TextUtils.isEmpty(newname)) {
                tv_name.setText(newname);
            }
        }
        return view;
    }
}
```

由以上代码我们可以看到，在适配器中实现"手机防盗"的标题替换的思路是：判断Item

对应的 id 是否为 0（即"手机防盗"对应的 Item），如果为 0，再判断 sp 中是否存入过用于替换的标题，如果存入了用于替换的标题，那么就用该标题，如果没有，那么就用数组中对应的标题——手机防盗。

运行测试：当成功进入"手机防盗"界面后，长按 Menu 键后，效果图如图 2-3 所示，当单击 Item 时，效果图如图 2-4 所示，然后在 EditText 文本框中输入"MP3"后，单击"确定"按钮，当再次进入程序主界面时，就可以看到修改后的标题生效了，界面效果图如图 2-5 所示。

图 2-3

图 2-4

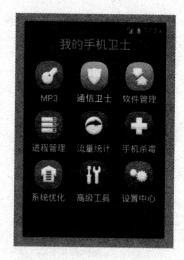
图 2-5

方案二的实现：实现外拨一个设定的电话号码后进入手机防盗界面。外拨一个电话时，系统会发送一个对应的广播。所以，我们可以通过广播来实现该功能。在包"com.guoshisp.mobilesafe.receiver"下创建一个 OutCallReceiver，同时继承 BroadcastReceiver，OutCallReceiver.java 的业务代码如下：

```java
package com.guoshisp.mobilesafe.receiver;
import android.content.BroadcastReceiver;
import android.content.Context;
import android.content.Intent;
import com.guoshisp.mobilesafe.LostProtectedActivity;
public class OutCallReceiver extends BroadcastReceiver {
    @Override
    public void onReceive(Context context, Intent intent) {
        //获取到广播发送来的数据结果
        String outnumber = getResultData();
        //设定拨号进入手机防盗的号码
        String enterPhoneBakNumber = "110";
        //判断设定的号码是否与广播过来的数据相同
        if (enterPhoneBakNumber.equals(outnumber)) {
            //进入手机防盗界面
```

```
            Intent lostIntent = new Intent(context, LostProtectedActivity.
                        class);
            //为手机防盗对应的Activity设置一个新的任务栈
            lostIntent.setFlags(Intent.FLAG_ACTIVITY_NEW_TASK);
            context.startActivity(lostIntent);
            //拦截掉外拨的电话号码，在拨号记录中不会显示该号码
            setResultData(null);
        }
    }
}
```

同时还需要在清单文件中为该对象配置一下对应的信息。

外拨电话的权限信息：

`<uses-permission android:name="android.permission.PROCESS_OUTGOING_CALLS"/>`

组件信息：

```
    <receiver android:name=".receiver.OutCallReceiver">
        <intent-filter android:priority="1000">
                <action android:name="android.intent.action.
                            NEW_OUTGOING_CALL"/>
        </intent-filter>
    </receiver>
```

对于配置组件信息需要注意的是：

（1）android:name=".receiver.OutCallReceiver"不能够配置为android:name=". OutCallReceiver"，因为"."是代码当前应用程序的包名，在清单文件中可以看到应用程序的包名为package="com.guoshisp.mobilesafe"，所以，如果我们配置为后者，当应用程序在加载时，将会找不到该组件。

（2）android:priority="1000"是设置广播接收的权限，对于有序广播来说，权限的范围是 -1000~1000，值越高，优先级就越高。接收者既可以终止广播意图的继续传播，也可以篡改广播的内容。

（3）`<action android:name="android.intent.action.NEW_OUTGOING_CALL"/>`

设置广播接收者的动作，这里设置的是外拨电话的动作。所以，如果外拨电话时，那么该广播接收者就会接收到该广播。

代码解析：

lostIntent.setFlags(Intent.FLAG_ACTIVITY_NEW_TASK)，因为广播本身是没有任务栈的，而Activity是需要放入任务栈中才能执行任务的，所以需要为Activity设置一个新的任务栈来执行自身的任务。

测试运行：当运行程序并退出后，我们在拨号器上输入110，那么界面会直接跳转至图2-2，当我们查看拨号记录时，拨号记录中已没有该号码的记录（因为我们在广播接收者中将数据拦截后设置为null了）。

2.3 实现手机防盗中的设置向导UI

在实现"设置向导"的UI设置前，先来简单复习一下Android中样式（style）的用法。

在Android中可以这样定义样式：在res/values/style.xml（文件名称可以是任意的，但是需要.xml文件类型）文件中添加以下内容：

```xml
<?xml verson="1.0" encoding="utf-8"?>
<resources>
    <style name="test"><!-- 为样式定义一个全局唯一的名字-->
        <item name="android:textSize">18px</item><!-- name 属性的值表示的是
        引用了该样式的View控件的属性，即View控件中的android:textSize属性 -->
        <item name="android:textCorlor">#0000CC</item>
    </style>
</resources>
```

然后，在layout文件夹下的布局文件中可以这样引用上面的Android样式：

```xml
<?xml verson="1.0" encoding="utf-8"?>
<LinearLayout xmlns:android="http://schemas.android.com/apk/res/android" ...>
    <TextView style="@style/test"<!-- 指定要引用的那个样式的名称即可引用-->
    ....../>
</LinearLayout>
```

<style>元素中有一个parent属性。这个属性可以让当前样式继承一个父样式，并且具有父样式的值。当然，如果父样式的值不符合要求，自己也可以对其进行修改（覆盖），例如：

```xml
<?xml version="1.0" encoding="utf-8"?>
<resources>
    <style name="test">
        <item name="android:textSize">18px</item><!-- name 属性的值表示的是
        引用了该样式的View控件的属性，即View控件中的android:textSize属性 -->
        <item name="android:textCorlor">#0000CC</item>
    </style>
    <style name="subtest" parent="@style/test"><!-- 使用parent属性继承了
        "test"样式，如果在同一个文件中，可以直接使用名称-->
        <item name="android:textSize">22px</item><!-- name 属性的值表示的是
        引用了该样式的View控件的属性，即View控件中的android:
        textSize属性，这里是覆盖了父方的属性android:textSize -->
    </style>
</resources>
```

设置向导UI界面分为4个，按照顺序，其界面预览图如图2-6、图2-7、图2-8、图2-9所示。

图 2-6

图 2-7

图 2-8

图 2-9

上面 4 张设置向导的图片是模拟器中的预览图，所以用于换行的 "\n" 在模拟器中没有生效，这也是模拟器的一个 bug，在真机上测试是可行的。首先，观察单张图片可以看出，图片中所要使用到的 View 有很多是相同的，只是内容不同罢了。例如，图 2-6 中 "您的手机防盗卫士："下面显示的 4 条数据只是显示的内容不同，其他部分完全一样。这时，我们可以将相同的部分提取出来，不同的部分自己定义。提取出相同的部分，我们可以使用样式来实现。再将 4 张图片放在一起进行观察，同样有很多相似之处，这时，我们同样可以使用样式，同时可以在样式中使用 "继承" 来简化布局代码。图 2-6 至图 2-9 对应的布局文件在 layout 下面，分别为 setup1.xml，setup2.xml，setup3.xml，setup4.xml。

setup1.xml 的布局文件代码为：

```
<?xml version="1.0" encoding="utf-8"?>
```

```xml
<LinearLayout xmlns:android="http://schemas.android.com/apk/res/android"
    android:layout_width="match_parent"
    android:layout_height="match_parent"
    android:orientation="vertical" >
    <TextView
        style="@style/text_title_style"
        android:layout_width="match_parent"
        android:layout_height="wrap_content"
        android:text="1.欢迎使用手机防盗" />
    <View style="@style/image_divideline_style" />
    <TextView
        style="@style/text_content_style"
        android:layout_width="match_parent"
        android:layout_height="wrap_content"
        android:layout_marginTop="8dip"
        android:text="您的手机防盗卫士:" />
    <LinearLayout
        android:layout_width="match_parent"
        android:layout_height="wrap_content"
        android:orientation="horizontal" >
        <ImageView
            style="@style/image_start_style"
            android:layout_width="wrap_content"
            android:layout_height="wrap_content" />
        <TextView
            style="@style/text_content_style"
            android:paddingTop="2dip"
            android:text="sim卡变更报警" />
    </LinearLayout>
    <LinearLayout
        android:layout_width="match_parent"
        android:layout_height="wrap_content"
        android:orientation="horizontal" >
        <ImageView style="@style/image_start_style" />
        <TextView
            style="@style/text_content_style"
            android:paddingTop="2dip"
            android:text="GPS追踪" />
    </LinearLayout>
    <LinearLayout
        android:layout_width="match_parent"
        android:layout_height="wrap_content"
        android:orientation="horizontal" >
        <ImageView style="@style/image_start_style" />
        <TextView
            style="@style/text_content_style"
```

```xml
            android:paddingTop="2dip"
            android:text="远程数据销毁" />
    </LinearLayout>
    <LinearLayout
        android:layout_width="match_parent"
        android:layout_height="wrap_content"
        android:orientation="horizontal" >
        <ImageView style="@style/image_start_style" />
        <TextView
            style="@style/text_content_style"
            android:paddingTop="2dip"
            android:text="远程锁屏" />
    </LinearLayout>
    <LinearLayout
        android:layout_width="match_parent"
        android:layout_height="wrap_content"
        android:gravity="center_horizontal"
        android:orientation="horizontal" >
        <ImageView style="@style/image_online_style" />
        <ImageView style="@style/image_offline_style" />
        <ImageView style="@style/image_offline_style" />
        <ImageView style="@style/image_offline_style" />
    </LinearLayout>
    <RelativeLayout
        android:layout_width="match_parent"
        android:layout_height="fill_parent" >
        <ImageView
            style="@style/image_logo_style"
            android:src="@drawable/setup1"
            />
        <Button style="@style/button_next_style"
            />
    </RelativeLayout>
</LinearLayout>
```

代码解析：

```xml
<RelativeLayout
    android:layout_width="match_parent"
    android:layout_height="fill_parent" >
    <ImageView
        style="@style/image_logo_style"
        android:src="@drawable/setup1"
        />
    <Button style="@style/button_next_style"
        />
</RelativeLayout>
```

该布局对应的图形是图 2-10 中的整个方框部分。

图 2-10

其中的 ImageView 和 Button 对应的样式为：

```xml
<style name="image_logo_style">
    <item name="android:layout_width">fill_parent</item>
    <item name="android:layout_height">fill_parent</item>
    <item name="android:scaleType">center</item>
</style>
<style name="button_next_style">
    <item name="android:layout_width">wrap_content</item>
    <item name="android:layout_height">wrap_content</item>
    <item name="android:layout_alignParentBottom">true</item>
    <item name="android:layout_alignParentRight">true</item>
    <item name="android:text">下一步</item>
    <item name="android:drawableRight">@drawable/next</item>
    <item name="android:onClick">next</item>
</style>
```

由样式可知：ImageView 的长和高都是充满状态，其中<item name="android:scaleType">center</item>用于使该 ImageView 在中央进行缩放显示，所以图片大小才没有充满整个方框。这样做的原因在于：我们希望在相对布局中显示方框中的图形位置，如果在相对布局的根节点中设置其中所有的子控件都居中显示，那么正常显示出 ImageView 是没有问题的，但当我们设置其中的 Button 在右下角显示时，相对布局中的居中显示的设置就会失去效果。

Setup2.xml 的布局文件代码为：

```xml
<?xml version="1.0" encoding="utf-8"?>
<LinearLayout xmlns:android="http://schemas.android.com/apk/res/android"
    android:layout_width="match_parent"
    android:layout_height="match_parent"
    android:orientation="vertical" >
```

```xml
<TextView
    style="@style/text_title_style"
    android:layout_width="match_parent"
    android:layout_height="wrap_content"
    android:text="2.手机卡的绑定" />
<View style="@style/image_divideline_style" />
<TextView
    style="@style/text_content_style"
    android:layout_width="match_parent"
    android:layout_height="wrap_content"
    android:layout_marginTop="8dip"
    android:text="一旦绑定sim卡\n,下次重启手机如果sim卡变化,就发送报警短信" />
<RelativeLayout
    android:id="@+id/rl_setup2_bind"
    android:layout_width="match_parent"
    android:layout_height="wrap_content"
    android:background="@drawable/item_bg" >
    <TextView
        style="@style/text_content_style"
        android:paddingTop="8dip"
        android:text="点击绑定sim卡" />
    <ImageView
        android:id="@+id/iv_setup2_bind_status"
        android:layout_width="wrap_content"
        android:layout_height="wrap_content"
        android:layout_alignParentRight="true"
        android:src="@drawable/switch_off_normal" />
</RelativeLayout>
<LinearLayout
    android:layout_width="match_parent"
    android:layout_height="wrap_content"
    android:gravity="center_horizontal"
    android:orientation="horizontal" >
    <ImageView style="@style/image_offline_style" />
    <ImageView style="@style/image_online_style" />
    <ImageView style="@style/image_offline_style" />
    <ImageView style="@style/image_offline_style" />
</LinearLayout>
<RelativeLayout
    android:layout_width="match_parent"
    android:layout_height="fill_parent" >
    <ImageView
        style="@style/image_logo_style"
        android:src="@drawable/bind" />
    <Button style="@style/button_next_style" />
    <Button style="@style/button_pre_style" />
</RelativeLayout>
```

</LinearLayout>

Setup3.xml 的布局文件代码为：

```xml
<?xml version="1.0" encoding="utf-8"?>
<LinearLayout xmlns:android="http://schemas.android.com/apk/res/android"
    android:layout_width="match_parent"
    android:layout_height="match_parent"
    android:orientation="vertical" >
    <TextView
        style="@style/text_title_style"
        android:layout_width="match_parent"
        android:layout_height="wrap_content"
        android:text="3.设置安全号码" />
    <View style="@style/image_divideline_style" />
    <TextView
        style="@style/text_content_style"
        android:layout_width="match_parent"
        android:layout_height="wrap_content"
        android:layout_marginTop="8dip"
        android:text="sim 卡更换后\n 报警短信会发送到安全号码上" />
    <EditText
        android:inputType="phone"
        android:layout_width="match_parent"
        android:layout_height="wrap_content"
        android:hint="请输入安全号码或者选择一个号码"
        android:id="@+id/et_setup3_number"
        />
    <Button
        android:layout_width="match_parent"
        android:layout_height="wrap_content"
        android:text="选择联系人"
        android:onClick="selectContact"
        />
    <LinearLayout
        android:layout_width="match_parent"
        android:layout_height="wrap_content"
        android:gravity="center_horizontal"
        android:orientation="horizontal" >
        <ImageView style="@style/image_offline_style" />
        <ImageView style="@style/image_offline_style" />
        <ImageView style="@style/image_online_style" />
        <ImageView style="@style/image_offline_style" />
    </LinearLayout>
    <RelativeLayout
        android:layout_width="match_parent"
        android:layout_height="fill_parent" >
        <ImageView
```

```xml
            style="@style/image_logo_style"
            android:src="@drawable/phone" />
        <Button style="@style/button_next_style" />
        <Button style="@style/button_pre_style" />
    </RelativeLayout>
</LinearLayout>
```

Setup4.xml 的布局文件代码为:

```xml
<?xml version="1.0" encoding="utf-8"?>
<LinearLayout xmlns:android="http://schemas.android.com/apk/res/android"
    android:layout_width="match_parent"
    android:layout_height="match_parent"
    android:orientation="vertical" >
    <TextView
        style="@style/text_title_style"
        android:layout_width="match_parent"
        android:layout_height="wrap_content"
        android:text="4.恭喜你,设置完成" />
     <View style="@style/image_divideline_style" />
    <TextView
        style="@style/text_content_style"
        android:layout_width="match_parent"
        android:layout_height="wrap_content"
        android:layout_marginTop="8dip"
        android:text="强烈建议您开启防盗保护" />
    <TextView
        android:clickable="true"
        android:onClick="activeDeviceAdmin"
        style="@style/text_content_style"
        android:layout_width="match_parent"
        android:layout_height="wrap_content"
        android:layout_marginTop="8dip"
        android:text="单击激活deviceadmin\n(激活后可以远程锁屏,清除数据)" />
    <CheckBox
        android:id="@+id/cb_setup4_protect"
        android:layout_width="match_parent"
        android:layout_height="wrap_content"
        android:text="防盗保护没有开启" />
    <LinearLayout
        android:layout_width="match_parent"
        android:layout_height="wrap_content"
        android:gravity="center_horizontal"
        android:orientation="horizontal" >
        <ImageView style="@style/image_offline_style" />
        <ImageView style="@style/image_offline_style" />
        <ImageView style="@style/image_offline_style" />
        <ImageView style="@style/image_online_style" />
```

```xml
        </LinearLayout>
        <RelativeLayout
            android:layout_width="match_parent"
            android:layout_height="fill_parent" >
            <ImageView
                style="@style/image_logo_style"
                android:src="@drawable/setup4" />
            <Button style="@style/button_next_style"
                android:text="设置完成"
                />
            <Button style="@style/button_pre_style" />
        </RelativeLayout>
</LinearLayout>
```

样式存放在 values 文件夹下的 style.xml 中,style.xml 文件代码为:

```xml
<?xml version="1.0" encoding="utf-8"?>
<resources>
    <style name="text_title_style">
        <item name="android:layout_width">match_parent</item>
        <item name="android:layout_height">wrap_content</item>
        <item name="android:textColor">#66ff00</item>
        <item name="android:textSize">28sp</item>
    </style>
    <style name="image_divideline_style">
        <item name="android:layout_width">fill_parent</item>
        <item name="android:layout_height">1dip</item>
        <item name="android:layout_marginTop">5dip</item>
        <item name="android:background">@drawable/devide_line</item>
    </style>
    <style name="image_start_style">
        <item name="android:layout_width">wrap_content</item>
        <item name="android:layout_height">wrap_content</item>
        <item name="android:src">@android:drawable/star_big_on</item>
    </style>
    <style name="image_online_style">
        <item name="android:layout_width">wrap_content</item>
        <item name="android:layout_height">wrap_content</item>
        <item name="android:paddingLeft">3dip</item>
        <item name="android:src">@android:drawable/presence_online</item>
    </style>
    <style name="image_offline_style">
        <item name="android:paddingLeft">3dip</item>
        <item name="android:layout_width">wrap_content</item>
        <item name="android:layout_height">wrap_content</item>
        <item name="android:src">@android:drawable/presence_invisible</item>
    </style>
    <style name="image_logo_style">
```

```xml
        <item name="android:layout_width">fill_parent</item>
        <item name="android:layout_height">fill_parent</item>
        <item name="android:scaleType">center</item>
    </style>
    <style name="button_next_style">
        <item name="android:layout_width">wrap_content</item>
        <item name="android:layout_height">wrap_content</item>
        <item name="android:layout_alignParentBottom">true</item>
        <item name="android:layout_alignParentRight">true</item>
        <item name="android:text">下一步</item>
        <item name="android:drawableRight">@drawable/next</item>
        <item name="android:onClick">next</item>
    </style>
    <style name="button_pre_style">
        <item name="android:layout_width">wrap_content</item>
        <item name="android:layout_height">wrap_content</item>
        <item name="android:layout_alignParentBottom">true</item>
        <item name="android:layout_alignParentLeft">true</item>
        <item name="android:text">上一步</item>
        <item name="android:onClick">pre</item>
        <item name="android:drawableLeft">@drawable/previous</item>
    </style>
    <style name="text_content_style" parent="@style/text_title_style">
        <item name="android:textSize">20sp</item>
    </style>
</resources>
```

以上是"设置向导"的 4 张 UI 图代码,当单击图 2-10 中的右下角的"设置完成"按钮时,会进入"设置向导"的设置结果的一个界面,如图 2-11 所示,同时,该界面也是当用户进行过设置向导后,以后再次成功进入"手机防盗"时要显示的界面。

图 2-11

图 2-11 对应的布局文件是 layout 文件下的 lost_protected.xml，布局文件代码如下：

```xml
<?xml version="1.0" encoding="utf-8"?>
<LinearLayout xmlns:android="http://schemas.android.com/apk/res/android"
    android:layout_width="match_parent"
    android:layout_height="match_parent"
    android:orientation="vertical" >
    <TextView
        style="@style/text_title_style"
        android:layout_width="match_parent"
        android:layout_height="wrap_content"
        android:gravity="center_horizontal"
        android:text="手机防盗" />
    <View style="@style/image_divideline_style" />
    <LinearLayout
        android:layout_width="match_parent"
        android:layout_height="wrap_content"
        android:layout_marginTop="8dip"
        android:orientation="horizontal" >
        <TextView
            style="@style/text_content_style"
            android:layout_width="wrap_content"
            android:text="安全号码为:" />
        <TextView
            android:id="@+id/tv_lost_protect_number"
            style="@style/text_content_style"
            android:layout_width="fill_parent"
            android:gravity="right"
            android:text="5556" />
    </LinearLayout>
    <View
        style="@style/image_divideline_style"
        android:background="@drawable/listview_devider" />
    <RelativeLayout
        android:id="@+id/rl_lost_protect_setting"
        android:layout_width="match_parent"
        android:layout_height="wrap_content" >
        <TextView
            style="@style/text_content_style"
            android:layout_width="wrap_content"
            android:paddingTop="12dip"
            android:text="防盗保护设置" />
        <CheckBox
            android:id="@+id/cb_lost_protect_setting"
            android:focusable="false"
            android:clickable="false"
            android:layout_width="wrap_content"
```

```xml
            android:layout_height="wrap_content"
            android:layout_alignParentRight="true"
            android:checked="false"
            android:text="防盗保护没有开启" />
    </RelativeLayout>
    <View
        style="@style/image_divideline_style"
        android:background="@drawable/listview_devider" />
    <TextView
        android:id="@+id/tv_lost_protect_reentry_setup"
        style="@style/text_content_style"
        android:paddingTop="10dip"
        android:text="重新进入设置向导" />
    <View
        style="@style/image_divideline_style"
        android:background="@drawable/listview_devider" />
    <TextView
        style="@style/text_content_style"
        android:paddingTop="10dip"
        android:text="手机防盗指令:\n#*location*#获取手机位置\n#*alarm*#播放报
                      警音乐\n#*locksrceen*#远程锁屏\n#*wipedata*#清除数据" />
</LinearLayout>
```

至此，设置向导的 UI 设置全部完成。

当设置完防盗密码后，第一次成功进入手机防盗时，应当弹出设置向导的第一个界面，依次单击界面中的"下一步"按钮，4 个界面应当依次展现出来。进入设置向导界面应该是在 LostProtectedActivity 中正确地输入密码后才能进入。

LostProtectedActivity.java 的完整业务代码见在线资源包中代码文本部分 2.3.doc。

当用户正确输入密码时，如果用户之前进行过设置向导，那么界面将会显示设置向导后的结果界面，即 lost_protected.xml 对应的界面。如果用户没有进行过设置向导，那么将执行 Intent intent = new Intent(this,Setup1Activity.class)代码，进入第一个设置向导所对应的界面。

Setup1Activity.java 中的业务代码为：

```java
package com.guoshisp.mobilesafe;
import android.app.Activity;
import android.content.Intent;
import android.os.Bundle;
import android.view.View;
public class Setup1Activity extends Activity {
    @Override
    protected void onCreate(Bundle savedInstanceState) {
        super.onCreate(savedInstanceState);
        //进入到设置向导的第一个界面
        setContentView(R.layout.setup1);
```

```
    }
    /**
     * 当单击设置向导的第一个界面中右下角按钮——下一步时所要执行的方法
     * 因为在该 Button 中设置有属性 android:onClick=next，在下面的代码中省去该解释
     * @param view
     */
    public void next(View view){
        Intent intent = new Intent(this,Setup2Activity.class);
        startActivity(intent);
        finish();
//Activity 切换时播放动画。参数一：界面进入时的动画效果；参数二：界面出去时的动画效果
        overridePendingTransition(R.anim.alpha_in, R.anim.alpha_out);
    }
}
```

代码解析：

overridePendingTransition(R.anim.alpha_in, R.anim.alpha_out)

它是 Activity 中的一个方法，执行 Activity 切换时要播放的动画。参数一：界面进入时的动画效果；参数二：界面出去时的动画效果。使用该函数还需要注意以下两点：它必需紧挨着 startActivity()或者 finish()函数之后调用；它只在 Android2.0 及以上版本上适用。alpha_in.xml 和 alpha_out.xml 都存放在 res/anim 目录下。alpha_in.xml 文件中的代码如下：

```xml
<?xml version="1.0" encoding="utf-8"?>
<alpha xmlns:android="http://schemas.android.com/apk/res/android"
    android:fromAlpha="0.0"
    android:toAlpha="1.0"
    android:duration="300"
    >
</alpha>
```

这是一个 Tween 动画的透明度改变的动画效果。其中，android:fromAlpha="0.0"表示完全透明，android:toAlpha="1.0"表示正常显示，android:duration="300"表示从完全透明到正常显示时需要的时间，单位是毫秒。

alpha_out.xml 文件中的代码如下：

```xml
<?xml version="1.0" encoding="utf-8"?>
<alpha xmlns:android="http://schemas.android.com/apk/res/android"
    android:fromAlpha="1.0"
    android:toAlpha="0.0"
    android:duration="300"
    >
</alpha>
```

该界面对应的测试效果图如图 2-12 所示。当单击该界面中的"下一步"按钮时，会执行 Intent intent = new Intent(this,Setup2Activity.class)代码。

Setup2Activity.java 中的业务代码为：

```java
package com.guoshisp.mobilesafe;
import android.app.Activity;
import android.content.Intent;
import android.content.SharedPreferences;
import android.content.SharedPreferences.Editor;
import android.os.Bundle;
import android.telephony.TelephonyManager;
import android.text.TextUtils;
import android.view.View;
import android.view.View.OnClickListener;
import android.widget.ImageView;
import android.widget.RelativeLayout;
public class Setup2Activity extends Activity implements OnClickListener {
    private RelativeLayout rl_setup2_bind;//"单击绑定sim卡"的父控件，该控件
    //中存在两个子控件，获取该控件的目的在于为其设置单击事件，便于单击该控件中的任何
    //一个控件都响应到单击事件
    private ImageView iv_setup2_bind_status;//rl_setup2_bind中的一个子控件，
    //用于显示sim卡是否被绑定时的不同状态
    private SharedPreferences sp;//用于保存sim卡是否被绑定的信息，以便程序下次
    //加载时使用
    @Override
    protected void onCreate(Bundle savedInstanceState) {
        super.onCreate(savedInstanceState);
        setContentView(R.layout.setup2);
        rl_setup2_bind =(RelativeLayout) findViewById(R.id.rl_setup2_
                    bind);
        rl_setup2_bind.setOnClickListener(this);
        iv_setup2_bind_status =(ImageView) findViewById(R.id.iv_setup2_
                        bind_status);
        sp = getSharedPreferences("config", MODE_PRIVATE);
        //初始化的逻辑——判断sim卡是否被绑定
        String simseral = sp.getString("simserial", "");
        if(TextUtils.isEmpty(simseral)){
iv_setup2_bind_status.setImageResource(R.drawable.switch_off_normal);
        }else{
iv_setup2_bind_status.setImageResource(R.drawable.switch_on_normal);
        }
    }
    /**
     * 在设置向导的第二个界面中单击"单击绑定sim卡"时执行的单击事件
     */
    public void onClick(View v) {
        switch (v.getId()) {
        case R.id.rl_setup2_bind:
```

```java
            //判断当前sim卡的状态
            String simseral = sp.getString("simserial", "");
            if(TextUtils.isEmpty(simseral)){//sim卡未绑定
                Editor editor = sp.edit();
                editor.putString("simserial", getSimSerial());
                editor.commit();
                //因为sim卡的状态是未被绑定,所以,单击条目后应该设置为绑定的状态
iv_setup2_bind_status.setImageResource(R.drawable.switch_on_normal);
            }else{
                Editor editor = sp.edit();
                editor.putString("simserial", "");
                editor.commit();
iv_setup2_bind_status.setImageResource(R.drawable.switch_off_normal);
            }
            break;
        }
    }
    /**
     * 获取手机的sim卡串号
     */
    private String getSimSerial(){
        //sim卡是与电话相关的,需要在清单文件中配置权限:<uses-permission
          android:name="android.permission.READ_PHONE_STATE" />
        TelephonyManager tm = (TelephonyManager) getSystemService
                        (TELEPHONY_SERVICE);
        //返回sim卡的串号
        return tm.getSimSerialNumber();
    }
    /**
     * 单击界面的右下角的"下一步"按钮所要执行的方法
     * @param view
     */
    public void next(View view){
        Intent intent = new Intent(this,Setup3Activity.class);
        startActivity(intent);
        finish();
//自定义一个平移的动画效果。参数一:界面进入时的动画效果;参数二:界面出去时的动画效果
        overridePendingTransition(R.anim.tran_in, R.anim.tran_out);
    }
    /**
     * 单击界面左下角的"上一步"按钮所要执行的方法
     * @param view
     */
    public void pre(View view){
        Intent intent = new Intent(this,Setup1Activity.class);
        startActivity(intent);
```

```
        finish();
        //自定义一个透明度变化的动画效果。参数一:界面进入时的动画效果;参数二:界面
          出去时的动画效果
        overridePendingTransition(R.anim.alpha_in, R.anim.alpha_out);
    }
}
```

代码解析:

overridePendingTransition(R.anim.tran_in, R.anim.tran_out)

设置向导进入第二个界面单击"下一步"按钮时,播放一个平移的动画。tran_in.xml 与 tran_out.xml 文件都存放在 res\anim 目录下。

说明:在进入第二个设置向导界面后,单击"下一步"按钮平移动画,单击"上一步"按钮播放透明度变化的动画。

tran_in.xml 文件中的代码为:

```
<?xml version="1.0" encoding="utf-8"?>
<translate xmlns:android="http://schemas.android.com/apk/res/android"
    android:fromYDelta="0"
    android:toYDelta="0"
    android:fromXDelta="100%p"
    android:toXDelta="0"
    android:duration="300"
    >
    </translate>
```

代码解析:

android:fromYDelta="0"与 android:toYDelta="0"表示界面在屏幕的 Y 轴方向的位置不发生改变。

android:fromXDelta="100%p"与 android:toXDelta="0"表示界面在屏幕上的 X 轴方向的位置是从屏幕的最右端平移到最左端。

android:duration="300"表示完成上面的动作所需要的时间,单位是毫秒。

tran_out.xml 文件中的代码为:

```
<?xml version="1.0" encoding="utf-8"?>
<translate xmlns:android="http://schemas.android.com/apk/res/android"
    android:fromYDelta="0"
    android:toYDelta="0"
    android:fromXDelta="0"
    android:toXDelta="-100%p"
    android:duration="300"
    >
    </translate>
```

该界面对应的测试效果图如图 2-13 所示,单击界面中的"单击绑定 sim 卡"后的效果图如图 2-16 所示。同时,文字换行也能够正确地显示。当单击"下一步"按钮时,执行代码 Intent intent = new Intent(this,Setup3Activity.class),Setup3Activity.java 中的业务代码如下:

```java
package com.guoshisp.mobilesafe;
import android.app.Activity;
import android.content.Intent;
import android.os.Bundle;
import android.view.View;
public class Setup3Activity extends Activity {
    @Override
    protected void onCreate(Bundle savedInstanceState) {
        super.onCreate(savedInstanceState);
        setContentView(R.layout.setup3);
    }
    public void next(View view){
        Intent intent = new Intent(this,Setup4Activity.class);
        startActivity(intent);
        finish();
//自定义一个平移的动画效果。参数一:界面进入时的动画效果;参数二:界面出去时的动画效果
        overridePendingTransition(R.anim.tran_in, R.anim.tran_out);
    }
    public void pre(View view){
        Intent intent = new Intent(this,Setup2Activity.class);
        startActivity(intent);
        finish();
        //自定义一个透明度变化的动画效果。参数一:界面进入时的动画效果;参数二:界面
            出去时的动画效果
        overridePendingTransition(R.anim.alpha_in, R.anim.alpha_out);
    }
}
```

该界面对应的测试效果图如图 2-14 所示,单击"下一步"按钮,执行代码 Intent intent = new Intent(this,Setup4Activity.class),Setup4Activity.java 对应的业务代码如下:

```java
package com.guoshisp.mobilesafe;
import android.app.Activity;
import android.app.AlertDialog;
import android.app.AlertDialog.Builder;
import android.app.admin.DevicePolicyManager;
import android.content.ComponentName;
import android.content.DialogInterface;
import android.content.DialogInterface.OnClickListener;
import android.content.Intent;
import android.content.SharedPreferences;
import android.content.SharedPreferences.Editor;
import android.os.Bundle;
```

```java
import android.view.View;
import android.widget.CheckBox;
import android.widget.CompoundButton;
import android.widget.CompoundButton.OnCheckedChangeListener;
import com.guoshisp.mobilesafe.receiver.MyAdmin;
public class Setup4Activity extends Activity {
    private CheckBox cb_setup4_protect;
    private SharedPreferences sp;
    @Override
    protected void onCreate(Bundle savedInstanceState) {
        super.onCreate(savedInstanceState);
        setContentView(R.layout.setup4);
        sp = getSharedPreferences("config", MODE_PRIVATE);
        cb_setup4_protect = (CheckBox)findViewById(R.id.cb_setup4_protect);
        //用于数据的回显。判断手机防盗是否开启，默认情况下没有开启
        boolean protecting = sp.getBoolean("protecting", false);
        cb_setup4_protect.setChecked(protecting);
        cb_setup4_protect.setOnCheckedChangeListener(new
            OnCheckedChangeListener() {
            public void onCheckedChanged(CompoundButton buttonView,
                                        boolean isChecked) {
                Editor editor = sp.edit();
                if(isChecked){
                    editor.putBoolean("protecting", true);
                    cb_setup4_protect.setText("防盗保护已经开启");
                }else{
                    cb_setup4_protect.setText("防盗保护没有开启");
                    editor.putBoolean("protecting", false);
                }
                editor.commit();
            }
        });
    }
    /**
     * 单击设置向导的第 4 个界面中的"设置完成"时所执行的方法,当执行该方法时,说明设
       置向导已经完成
     * @param view
     */
    public void next(View view){
        if(!cb_setup4_protect.isChecked()){//如果防盗保护没有开启，弹出一个对
                                             话框提示开启保护
            AlertDialog.Builder builder = new Builder(this);
            builder.setTitle("温馨提示");
            builder.setMessage("手机防盗极大地保护了你的手机安全,强烈建议开启!");
            builder.setPositiveButton("开启", new OnClickListener() {
```

```java
            public void onClick(DialogInterface dialog, int which) {
                //将防盗保护开启
                cb_setup4_protect.setChecked(true);
        //设置向导已经完成,在用户下次进入时判断,值为true,说明已经进行过设置向导
                Editor editor = sp.edit();
                editor.putBoolean("issetup", true);
                editor.commit();
            }
        });
        builder.setNegativeButton("取消", new OnClickListener() {
            public void onClick(DialogInterface dialog, int which) {
                finish();
        //设置向导已经完成,在用户下次进入时判断,值为true,说明已经进行过设置向导
                Editor editor = sp.edit();
                editor.putBoolean("issetup", true);
                editor.commit();
            }
        });
        builder.create().show();
        return ;
    }
    //设置向导已经完成,在用户下次进入时判断,值为true,说明已经进行过设置向导
    Editor editor = sp.edit();
    editor.putBoolean("issetup", true);
    editor.commit();
    Intent intent = new Intent(this,LostProtectedActivity.class);
    startActivity(intent);
    finish();
    //自定义一个平移的动画效果。参数一:界面进入时的动画效果;参数二:界面出去时
        的动画效果
    overridePendingTransition(R.anim.tran_in, R.anim.tran_out);
}
public void pre(View view){
    Intent intent = new Intent(this,Setup3Activity.class);
    startActivity(intent);
    finish();
    //自定义一个透明度变化的动画效果。参数一:界面进入时的动画效果;参数二:界面
        出去时的动画效果
    overridePendingTransition(R.anim.alpha_in, R.anim.alpha_out);
    }
}
```

该界面对应的测试效果图如图 2-15 所示,单击"设置完成"时,如果防盗保护没有开启的话,会弹出一个窗体提示我们开启防盗保护,如图 2-17 所示。

第 2 章 手机防盗模块的设计

图 2-12

图 2-13

图 2-14

图 2-15

图 2-16

图 2-17

2.4 获取联系人的数据与完成设置向导逻辑

在设置向导的第三个界面（图 2-14）有一个"选择联系人"来完成安全号码绑定的操作。Android 手机中的联系人保存在 data\data\com.android.providers.contacts\databases 目录下的 contacts2.db 文件中。我们将其导出到桌面上，然后通过 SQLite Expert 工具来打开该文件。因为模拟器中的联系人列表中没有任何联系人，所以我们在 SQLite Expert 工具中查看 data 表和 raw_contacts 表时，表中的数据为空。此时，在联系人列表中添加联系人姓名：ZhangSan（Android2.2 模拟器中是将姓氏和名字分开的，姓氏：Zhang；名字：San），电话号码：119。此时，我们再次将联系人数据文件导到桌面上，然后在 SQLite Expert 工具中单击刷新，接着用鼠标左键单击一下 raw_contacts 表，在右边我们单击上面的"Data"查看数据时发现多出了一条数据，如图 2-18 所示。在表中，可以看到有"_id"和"contact_id"。下面介绍 raw_contacts 表、data 表、mimetypes 表以及 raw_contacts 表和 data 表是如何建立联系的。

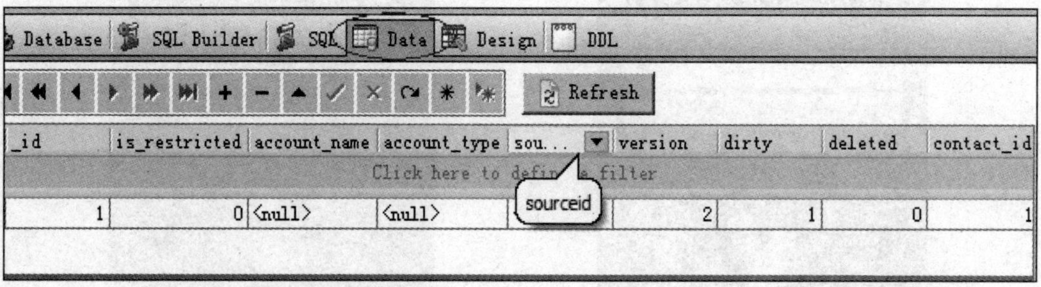

图 2-18

raw_contacts 表：用来存放联系人的 id，有多少联系人，就会在该表中出现多少条数据。

data 表：用来存放联系人的具体数据，如图 2-19 所示。

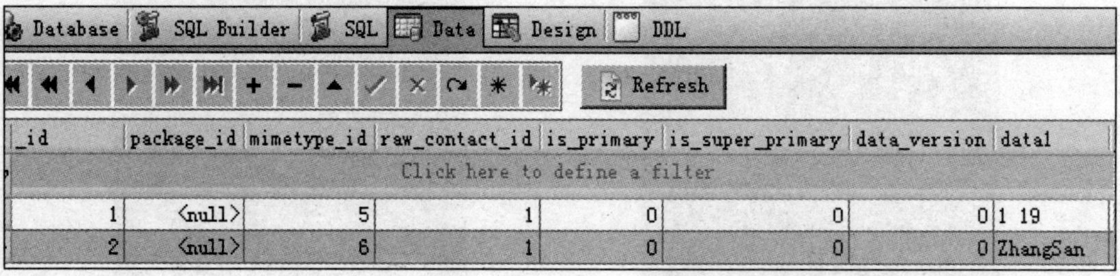

图 2-19

raw_contacts 表中联系人的 id 和 data 表中对应的具体数据是通过 data 表中的外键 raw_contact_id 建立映射关系的。具体映射过程：首先会在 raw_contacts 表中生成一条记录为 contact_id，然后会在 data 表中根据 raw_contact_id 来标识出对应的联系人。

mimetypes 表：保存数据类型。在 data 表中"data1"列下面有两条数据，分别为 119 和 ZhangSan，这两条数据所对应的 mimetype_id 分别为 5 和 6。我们打开 mimetypes 表，查看 5 和 6，如图 2-20 所示。

为了验证我们的解析，可以在联系人中多添加一条联系人，并且在创建联系人的信息时多添加几条信息，例如，电子邮件信息。读者有兴趣的话，可以动手做实验。

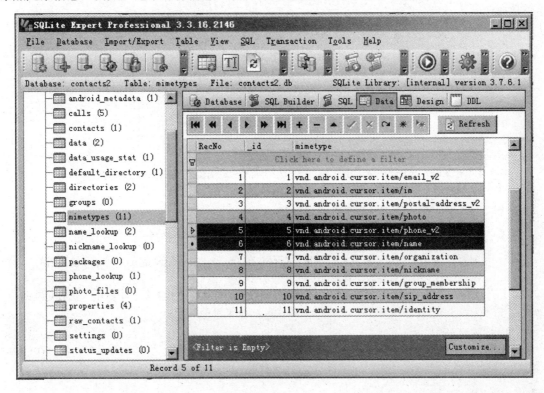

图 2-20

获取联系人的操作位于在设置向导的第三个界面——Setup3Activity 对应的界面中，至此，Setup3Activity.java 的全部业务代码为：

```
package com.guoshisp.mobilesafe;
import android.app.Activity;
import android.content.Intent;
import android.content.SharedPreferences;
import android.content.SharedPreferences.Editor;
import android.os.Bundle;
import android.text.TextUtils;
import android.view.View;
import android.widget.EditText;
import android.widget.Toast;
public class Setup3Activity extends Activity {
    private EditText et_setup3_number;//设置绑定的安全号码
    private SharedPreferences sp;//用于存储安全号码及安全号码的回显
```

```java
@Override
protected void onCreate(Bundle savedInstanceState) {
    super.onCreate(savedInstanceState);
    setContentView(R.layout.setup3);
    et_setup3_number = (EditText) findViewById(R.id.et_setup3_number);
    sp = getSharedPreferences("config", MODE_PRIVATE);
    //数据的回显。如果没有保存过安全号码,回显的是空
    String number = sp.getString("safemuber", "");
    et_setup3_number.setText(number);
}
/**
 * 在第三个界面的"选择联系人"按钮上设置有属性:android:onClick=
   "selectContact",所以,当单击"选择联系人"时会执行该方法
 * @param view
 */
public void selectContact(View view){
    Intent intent = new Intent(this,SelectContactActivity.class);
    //激活一个带返回值的Activity。参数二:请求码
    startActivityForResult(intent, 0);
}
/**
 * 被激活的Activity将返回的结果数据存放在Intent中,这里的Intent和被激活的
   Activity返回数据时所使用的是同一个Intent
 * 注意:如果希望数据能够正常返回,Activity的启动模式不能设置为singletask模式
 */
@Override
protected void onActivityResult(int requestCode, int resultCode, Intent
                                data) {
    if(data!=null){
        //获取到返回的数据
        String number = data.getStringExtra("number");
        //将返回的数据显示在EditText中
        et_setup3_number.setText(number);
    }
    super.onActivityResult(requestCode, resultCode, data);
}
//单击"下一步"执行的方法
public void next(View view){
    String number = et_setup3_number.getText().toString().trim();
    if(TextUtils.isEmpty(number)){
        Toast.makeText(this, "安全号码不能为空", 0).show();
        return;
    }
    //将EditText中的安全号码持久化,也方便数据的回显
    Editor editor = sp.edit();
    editor.putString("safemuber", number);
```

```
            editor.commit();
            Intent intent = new Intent(this,Setup4Activity.class);
            startActivity(intent);
            finish();
//自定义一个平移的动画效果。参数一：界面进入时的动画效果；参数二：界面出去时的动画效果
            overridePendingTransition(R.anim.tran_in, R.anim.tran_out);
        }
        //单击"上一步"执行的方法
        public void pre(View view){
            Intent intent = new Intent(this,Setup2Activity.class);
            startActivity(intent);
            finish();
            //自定义一个透明度变化的动画效果。参数一：界面进入时的动画效果；参数二：界面
               出去时的动画效果
            overridePendingTransition(R.anim.alpha_in, R.anim.alpha_out);
        }
}
```

代码解析：

（1）public void selectContact(View view){

　　Intent intent = new Intent(this,SelectContactActivity.class);
　　//激活一个带返回值的Activity。参数二：请求码
　　startActivityForResult(intent, 0);

　　}

当单击"选择联系人"按钮时执行该方法。执行 startActivityForResult(intent, 0)时激活了 SelectContactActivity，同时要求该 Activity 返回数据。

（2）@Override

protected void onActivityResult(int requestCode, int resultCode, Intent data) {
　　if(data!=null){
　　　　//获取到返回的数据
　　　　String number = data.getStringExtra("number");
　　　　//将返回的数据显示在 EditText 中
　　　　et_setup3_number.setText(number);
　　}
　　super.onActivityResult(requestCode, resultCode, data);
}

当被要求返回数据的 SelectContactActivity 执行 setResult(0, data)时就执行该方法，该方法用于接收 SelectContactActivity 所返回的数据。

SelectContactActivity.java 中的业务代码为：

```
package com.guoshisp.mobilesafe;
```

```java
import java.util.List;
import android.app.Activity;
import android.content.Intent;
import android.graphics.Color;
import android.os.Bundle;
import android.view.View;
import android.view.ViewGroup;
import android.widget.AdapterView;
import android.widget.AdapterView.OnItemClickListener;
import android.widget.BaseAdapter;
import android.widget.ListView;
import android.widget.TextView;
import com.guoshisp.mobilesafe.domain.ContactInfo;
import com.guoshisp.mobilesafe.engine.ContactInfoProvider;
public class SelectContactActivity extends Activity {
    private ListView lv_select_contact;//用于展现联系人的列表
    private ContactInfoProvider provider;//获取手机联系人的对象
    private List<ContactInfo> infos;//接收获取到的所有联系人
    @Override
    protected void onCreate(Bundle savedInstanceState) {
        super.onCreate(savedInstanceState);
        setContentView(R.layout.select_contact);
        lv_select_contact = (ListView) findViewById(R.id.lv_select_
                        contact);
        provider = new ContactInfoProvider(this);
        infos = provider.getContactInfos();
        //为 lv_select_contact 设置一个数据适配器，用于将所有联系人展现到界面上
        lv_select_contact.setAdapter(new ContactAdapter());
        //为 lv_select_contact 中的 item 设置监听
        lv_select_contact.setOnItemClickListener(new OnItemClickListener() {
            public void onItemClick(AdapterView<?> parent, View view,
                int position, long id) {
                //获取到单击 item 对应的联系人的信息对象
                ContactInfo info = (ContactInfo) lv_select_contact.
                                getItemAtPosition(position);
                //获取到该联系人的号码
                String number = info.getPhone();
                //将该联系人的号码返回给激活当前 Activity 的 Activity
                Intent data = new Intent();
                //将数据存入，用于返回给 Activity
                data.putExtra("number", number);
                //返回数据，参数一：返回结果码；参数二：返回数据
                setResult(0, data);
                //关闭当前的 Activity
                finish();
            }
```

```java
        });
    }
    /**
     * 展现所有联系人
     * @author Administrator
     *
     */
    private class ContactAdapter extends BaseAdapter{
        public int getCount() {
            return infos.size();
        }
        public Object getItem(int position) {
            return infos.get(position);
        }
        public long getItemId(int position) {
            return position;
        }
        public View getView(int position, View convertView, ViewGroup
                            parent) {
            ContactInfo info = infos.get(position);
            TextView tv = new TextView(getApplicationContext());
            tv.setTextSize(24);
            tv.setTextColor(Color.WHITE);
            tv.setText(info.getName()+"\n"+info.getPhone());
            return tv;
        }
    }
}
```

在 layout 下对应的 select_contact.xml 布局文件为：

```xml
<?xml version="1.0" encoding="utf-8"?>
<LinearLayout xmlns:android="http://schemas.android.com/apk/res/android"
    android:layout_width="fill_parent"
    android:layout_height="fill_parent"
    android:orientation="vertical" >
    <TextView
        android:layout_width="wrap_content"
        android:layout_height="wrap_content"
        android:layout_gravity="center_horizontal"
        android:text="选择联系人"
        android:textColor="#66ff00"
        android:textSize="28sp" />
    <View
        android:layout_width="fill_parent"
        android:layout_height="1dip"
        android:layout_marginTop="5dip"
```

```xml
        android:background="@drawable/devide_line" />
    <ListView
        android:id="@+id/lv_select_contact"
        android:layout_width="fill_parent"
        android:layout_height="fill_parent"
        android:layout_marginTop="5dip" >
    </ListView>
</LinearLayout>
```

获取手机联系人的对象 ContactInfoProvider 存放在包"com.guoshisp.mobilesafe.engine"下，ContactInfoProvider.java 代码如下：

```java
package com.guoshisp.mobilesafe.engine;
import java.util.ArrayList;
import java.util.List;
import android.content.Context;
import android.database.Cursor;
import android.net.Uri;
import com.guoshisp.mobilesafe.domain.ContactInfo;
public class ContactInfoProvider {
    private Context context;
    public ContactInfoProvider(Context context) {
        this.context = context;
    }
    /**
     * 返回所有的联系人的信息
     *
     * @return
     */
    public List<ContactInfo> getContactInfos() {
        //将所有联系人存入该集合
        List<ContactInfo> infos = new ArrayList<ContactInfo>();
        //获取 raw_contacts 表所对应的 Uri
        Uri uri = Uri.parse("content://com.android.contacts/raw_contacts");
        //获取 data 表所对应的 Uri
        Uri datauri = Uri.parse("content://com.android.contacts/data");
        //参数二：所要查询的列，即联系人的 id。获取一个查询数据库所返回的结果集
        Cursor cursor = context.getContentResolver().query(uri,
                new String[] { "contact_id" }, null, null, null);
        while (cursor.moveToNext()) {//移动游标
            //因为只需要查询一列数据——联系人的 id，所以我们传入 0
            String id = cursor.getString(0);
            //用于封装每个联系人的具体信息
            ContactInfo info = new ContactInfo();
            //得到 id 后，我们通过该 id 来查询 data 表中的联系人的具体数据（data 表中的 data1 中的数据）。参数二：null，会将所有的列返回回来
```

```
            //参数三：选择条件，返回一个在data表中查询后的结果集
            Cursor dataCursor = context.getContentResolver().query(datauri,
                    null, "raw_contact_id=?", new String[] { id }, null);
            while (dataCursor.moveToNext()) {
            //dataCursor.getString(dataCursor.getColumnIndex("mimetype"))
                获取data1列中具体数据的数据类型，这里判断的是联系人的姓名
                if ("vnd.android.cursor.item/name".equals(dataCursor
                        .getString(dataCursor.getColumnIndex
                            ("mimetype")))){
            //dataCursor.getString(dataCursor.getColumnIndex("data1"))获
                取data1列中的联系人的具体数据
                    info.setName(dataCursor.getString(dataCursor
                        .getColumnIndex("data1")));
                } else if ("vnd.android.cursor.item/phone_v2".equals
                    (dataCursor
                    .getString(dataCursor.getColumnIndex
                        ("mimetype")))){ //数据类型是否是手机号码
                    info.setPhone(dataCursor.getString(dataCursor
                        .getColumnIndex("data1")));
                }
            }
            //每查询一个联系人后就将其添加到集合中
            infos.add(info);
            info = null;
            dataCursor.close();//关闭结果集
        }
        cursor.close();
        return infos;
    }
}
```

封装每个联系人的是数据的ContactInfo对象存放在包"com.guoshisp.mobilesafe.domain"下，ContactInfo.java代码为：

```
package com.guoshisp.mobilesafe.domain;
public class ContactInfo {
    private String name;
    private String phone;
    public String getName() {
        return name;
    }
    public void setName(String name) {
        this.name = name;
    }
    public String getPhone() {
        return phone;
```

```
}
public void setPhone(String phone) {
    this.phone = phone;
}
}
```

另外，不仅应该在清单文件中为每个 Activity 配置好对应的信息，而且还要加入获取手机串号和操作联系人的权限信息：

```
<uses-permission android:name="android.permission.READ_PHONE_STATE" />
<uses-permission android:name="android.permission.READ_CONTACTS" />
```

测试运行：在模拟器的联系人中多添加几个联系人，当单击设置向导中的第三个界面中的"选择联系人"按钮时，界面如图 2-21 所示，选择"ZhangSan"后，设置向导的第三个界面如图 2-22 所示。

至此，选择联系人的操作全部完成。

图 2-21　　　　　　　　　　　图 2-22

2.5　实现手机防盗指令

手机防盗的一个最核心的功能是绑定手机的 sim 卡，当手机的 sim 卡发生改变时，需要给我们的安全号码发送信息。对于 Android 系统的手机，当 sim 卡发生改变时，系统会重新启动，然后重新识别 sim 卡。所以，我们可以在手机重启时，获取当前 sim 卡的串号是否与之前绑定的 sim 卡的串号相同。如果不同，则向之前绑定的安全号码发送短信。我们知道，手机重启时会发出一个广播，我们通过广播接收者来完成对 sim 卡串号的判断。在包"com.guoshisp.mobilesafe.receiver"下创建出 BootCompleteReceiver 对象，用于接收手机重启的广播事件。

BootCompleteReceiver.java 代码如下：

```java
package com.guoshisp.mobilesafe.receiver;
import android.content.BroadcastReceiver;
import android.content.Context;
import android.content.Intent;
import android.content.SharedPreferences;
import android.telephony.SmsManager;
import android.telephony.TelephonyManager;
import android.util.Log;
public class BootCompleteReceiver extends BroadcastReceiver {
    private static final String TAG = "BootCompleteReceiver";
    @Override
    public void onReceive(Context context, Intent intent) {
        Log.i(TAG,"手机重启了");
        SharedPreferences sp = context.getSharedPreferences("config",
                    Context.MODE_PRIVATE);
        boolean protecting = sp.getBoolean("protecting", false);
        if(protecting){
            String safemuber = sp.getString("safemuber", "");
            //判断当前手机的sim卡和绑定的sim是否一致
            TelephonyManager tm = (TelephonyManager) context.getSystemService
                            (Context.TELEPHONY_SERVICE);
            String realsim = tm.getSimSerialNumber();
            String savedSim = sp.getString("simserial", "");
            if(!savedSim.equals(realsim)){
                //发送报警短信
                Log.i(TAG,"发送短信");
                SmsManager smsManager = SmsManager.getDefault();
                smsManager.sendTextMessage(safemuber, null, "sim card
                                changed", null, null);
            }
        }
    }
}
```

为其在清单文件中配置相应的信息：

```xml
<receiver android:name=".receiver.BootCompleteReceiver" >
        <intent-filter android:priority="1000" >
            <action android:name="android.intent.action.BOOT_COMPLETED" />
        </intent-filter>
</receiver>
```

其中的<action android:name="android.intent.action.BOOT_COMPLETED" />是开机启动的动作，用于接收开机启动的广播。

还需要配置接收开机自动启动广播的权限和发送短信的权限：

```xml
<uses-permission android:name="android.permission.RECEIVE_BOOT_COMPLETED" />
<uses-permission android:name="android.permission.SEND_SMS" />
```

测试：将安全号码修改为 5556。由于模拟器不能更改 sim 卡，可以将 sp 中的 sim 卡串号修改一下，创建并启动 5556 模拟器，然后重启一下 5554 模拟器。如果测试成功，5556 模拟器应该接收一条内容为"sim card changed"的信息（模拟器不支持中文）。测试结果如图 2-23 所示，其中，1555 表示的模拟器的主机号，521 表示的是模拟器的端口号。

当接收到报警短信后（也即获取到了对方的手机号码），可以向手机发送一些指令：播放报警音乐、锁屏、获取位置信息、清除数据恢复至出厂设置。手机接收到一条短信，同样也是接收到了一个广播。所以可以通过一个广播来实现对接收到的信息的处理。创建一个广播接收者 SmsReceiver。

图 2-23

SmsReceiver.java 中的业务代码如下：

```java
package com.guoshisp.mobilesafe.receiver;
import android.app.admin.DevicePolicyManager;
import android.content.BroadcastReceiver;
import android.content.ComponentName;
import android.content.Context;
import android.content.Intent;
import android.content.SharedPreferences;
import android.media.MediaPlayer;
import android.telephony.SmsManager;
import android.telephony.SmsMessage;
import android.text.TextUtils;
import android.util.Log;
import com.guoshisp.mobilesafe.R;
import com.guoshisp.mobilesafe.engine.GPSInfoProvider;
```

```java
public class SmsReceiver extends BroadcastReceiver {
    private static final String TAG = "SmsReceiver";
    private SharedPreferences sp;
    @Override
    public void onReceive(Context context, Intent intent) {
        Log.i(TAG,"短信到来了");
        sp = context.getSharedPreferences("config", Context.MODE_PRIVATE);
        String safenumber = sp.getString("safemuber", "");
        //获取短信中的内容。系统接收到一个信息广播时，会将接收到的信息存放到pdus数组中
        Object[] objs = (Object[]) intent.getExtras().get("pdus");
        //获取手机设备管理器
        DevicePolicyManager dm = (DevicePolicyManager) context.getSystemService
                        (Context.DEVICE_POLICY_SERVICE);
        //创建一个与MyAdmin相关联的组件
        ComponentName mAdminName = new ComponentName(context, MyAdmin.class);
        //遍历出信息中的所有内容
        for(Object obj : objs){
            SmsMessage smsMessage = SmsMessage.createFromPdu((byte[]) obj);
            //获取发件人的地址
            String sender = smsMessage.getOriginatingAddress();
            //获取短信信息内容
            String body = smsMessage.getMessageBody();
                if("#*location*#".equals(body)){
                Log.i(TAG,"发送位置信息");
                //获取当前的位置
                String lastlocation = GPSInfoProvider.getInstance(context).
                                getLocation();
                if(!TextUtils.isEmpty(lastlocation)){
                    //得到信息管理器
                    SmsManager smsManager = SmsManager.getDefault();
                    //向安全号码发送当前的位置信息
                    smsManager.sendTextMessage(safenumber, null,
                                    lastlocation, null, null);
                }
                abortBroadcast();
            }else if("#*alarm*#".equals(body)){
                Log.i(TAG,"播放报警音乐");
                //得到音频播放器
                MediaPlayer player = MediaPlayer.create(context, R.raw.
                                ylzs);//res\raw\ylzs.mp3
                //即使手机是静音模式也有音乐的声音
                player.setVolume(1.0f, 1.0f);
                //开始播放音乐
                player.start();
                //终止发送过来的信息，在本地查看不到该信息
                abortBroadcast();
```

```
            }else if("#*wipedata*#".equals(body)){
                Log.i(TAG,"清除数据");
                //判断设备的管理员权限是否被激活。只有被激活后，才可以执行锁频、清
                  除数据、重置出厂设置（模拟器不支持该操作）等操作
                if(dm.isAdminActive(mAdminName)){
                    dm.wipeData(0);//清除设备中的数据，手机会自动重启
                }
                abortBroadcast();
            }else if("#*lockscreen*#".equals(body)){
                Log.i(TAG,"远程锁屏");
                if(dm.isAdminActive(mAdminName)){
                    dm.resetPassword("123", 0);//屏幕解锁时需要的解锁密码为123
                    dm.lockNow();
                }
                abortBroadcast();
            }
        }
    }
}
```

在清单文件中配置组件信息及相关权限：

```
<receiver android:name=".receiver.SmsReceiver" >
        <intent-filter android:priority="1000" >
            <action android:name="android.provider.Telephony.SMS_RECEIVED" />
        </intent-filter>
</receiver>
<uses-permission android:name="android.permission.RECEIVE_SMS" />
    <uses-permission android:name="android.permission.ACCESS_MOCK_LOCATION" />
    <uses-permission android:name="android.permission.ACCESS_FINE_LOCATION" />
    <uses-permission android:name="android.permission.ACCESS_COARSE_LOCATION" />
```

GPSInfoProvider 在包"com.guoshisp.mobilesafe.engine"下，GPSInfoProvider.java 代码如下：

```
package com.guoshisp.mobilesafe.engine;
import android.content.Context;
import android.content.SharedPreferences;
import android.content.SharedPreferences.Editor;
import android.location.Criteria;
import android.location.Location;
import android.location.LocationListener;
import android.location.LocationManager;
import android.os.Bundle;
public class GPSInfoProvider {
    private static GPSInfoProvider mGPSInfoProvider;
    private static LocationManager lm;//位置管理器
    private static MyListener listener;//位置变化的监听器，监听动作比较耗电
```

```java
private static SharedPreferences  sp;//持久化位置的信息(经纬度)
//私有化构造方法,做成单例模式。目的是减少往系统服务注册监听,避免程序挂掉,减少耗电量
private GPSInfoProvider(){
}
public synchronized static GPSInfoProvider getInstance(Context context){
    if(mGPSInfoProvider==null){
        mGPSInfoProvider = new GPSInfoProvider();
        //获取位置管理器
        lm = (LocationManager) context.getSystemService(Context.
            LOCATION_SERVICE);
        //获取查询地理位置的查询条件对象(内部是一个Map集合)
        Criteria criteria = new Criteria();
        //设置精度,这里传递的是最精准的精确度
        criteria.setAccuracy(Criteria.ACCURACY_FINE);
        //gps定位是否允许产生开销(true表示允许,如好用流量)
        criteria.setCostAllowed(true);
        //手机的功耗消耗情况(实时定位时,应该设置为高)
        criteria.setPowerRequirement(Criteria.POWER_HIGH);
        //获取海拔信息
        criteria.setAltitudeRequired(true);
        //对手机的移动的速度是否敏感
        criteria.setSpeedRequired(true);
        //获取到当前手机最好用的位置提供者:参数一:查询的选择条件;参数二:传递
            为true时,表示只有可用的位置提供者时才会被返回
        String provider = lm.getBestProvider(criteria, true);
        //System.out.println(provider);
        listener = new GPSInfoProvider().new MyListener();
        //调用更新位置方法。参数一:位置提供者;参数二:最短的更新位置信息时间(最
            好大于60000(一分钟));参数三:最短通知距离;参数四:位置改变时的监
            听对象
        lm.requestLocationUpdates(provider, 60000, 100, listener);
        //在Sdcard对应的包中创建一个config.xml文件,文件的操作类型设置为PRIVATE
        sp = context.getSharedPreferences("config", Context.MODE_
            PRIVATE);
    }
    return mGPSInfoProvider;
}
/**
 * 取消位置的监听
 */
public void stopLinsten(){
    lm.removeUpdates(listener);
    listener = null;
}
protected class MyListener implements LocationListener {
```

```java
    /**
     * 当手机的位置发生改变时调用的方法
     */
public void onLocationChanged(Location location) {
String latitude = "latitude :" + location.getLatitude(); //纬度
String longitude = "longitude: " + location.getLongitude(); //经度
String meter = "accuracy :" + location.getAccuracy();//精确度
System.out.println(latitude + "-" + longitude + "-" + meter);
    Editor editor = sp.edit();
     editor.putString("last_location", latitude + "-" + longitude + "-"
            + meter);
        editor.commit();
    }
    /**
     * 当位置提供者状态发生改变时调用的方法
     */
    public void onStatusChanged(String provider, int status, Bundle
                                extras) {
    }
    /**
     * 当某个位置提供者可用时
     */
    public void onProviderEnabled(String provider) {
    }
    /**
     * 当某个位置提供者不可用时
     */
    public void onProviderDisabled(String provider) {
    }
}
/**
 * 获取手机的位置
 * @return
 */
public String getLocation(){
    return sp.getString("last_location", "");
}
}
```

以上可以完成位置信息的获取，播放报警音乐（当离手机比较近时比较实用）。如果需要进行远程锁频和清空数据恢复到出厂设置，则先需要为当前应用程序激活设备的超级管理员权限。在设置向导的第 4 个界面有一个 TextView 为"单击激活 deviceadmin\n（激活后可以远程锁屏，清除数据）"的条目，该 View 设置有属性：android:clickable="true"和 android:onClick="activeDeviceAdmin"。

至此，Setup4Activity.java 中的全部业务代码如下：

```java
package com.guoshisp.mobilesafe;
import android.app.Activity;
import android.app.AlertDialog;
import android.app.AlertDialog.Builder;
import android.app.admin.DevicePolicyManager;
import android.content.ComponentName;
import android.content.DialogInterface;
import android.content.DialogInterface.OnClickListener;
import android.content.Intent;
import android.content.SharedPreferences;
import android.content.SharedPreferences.Editor;
import android.os.Bundle;
import android.view.View;
import android.widget.CheckBox;
import android.widget.CompoundButton;
import android.widget.CompoundButton.OnCheckedChangeListener;
import com.guoshisp.mobilesafe.receiver.MyAdmin;
public class Setup4Activity extends Activity {
    private CheckBox cb_setup4_protect;
    private SharedPreferences sp;
    @Override
    protected void onCreate(Bundle savedInstanceState) {
        super.onCreate(savedInstanceState);
        setContentView(R.layout.setup4);
        sp = getSharedPreferences("config", MODE_PRIVATE);
        cb_setup4_protect = (CheckBox)findViewById(R.id.cb_setup4_protect);
        //用于数据的回显。判断手机防盗是否开启，默认情况下没有开启
        boolean protecting = sp.getBoolean("protecting", false);
        cb_setup4_protect.setChecked(protecting);

        cb_setup4_protect.setOnCheckedChangeListener(new
                OnCheckedChangeListener() {
            public void onCheckedChanged(CompoundButton buttonView,
                                        boolean isChecked) {
                Editor editor = sp.edit();
                if(isChecked){
                    editor.putBoolean("protecting", true);
                    cb_setup4_protect.setText("防盗保护已经开启");
                }else{
                    cb_setup4_protect.setText("防盗保护没有开启");
                    editor.putBoolean("protecting", false);
                }
                editor.commit();
```

```java
            }
        });
    }
    /**
     * 当单击设置向导中的第4个界面中的"单击激活deviceadmin…"时所执行的方法
     * 激活手机的设备管理员权限。激活后，可以执行远程锁屏、清除数据恢复至出厂设置
     * @param view
     */
    public void activeDeviceAdmin(View view){
        //创建一个与MyAdmin相关联的组件
        ComponentName mAdminName = new ComponentName(this, MyAdmin.class);
        //获取手机设备管理器
        DevicePolicyManager dm = (DevicePolicyManager) getSystemService
                        (DEVICE_POLICY_SERVICE);
        //判断组件是否已经获取超级管理员的权限
        if (!dm.isAdminActive(mAdminName)) {
            Intent intent = new Intent(
                DevicePolicyManager.ACTION_ADD_DEVICE_ADMIN);
            //将组件的超级管理员权限激活
            intent.putExtra(DevicePolicyManager.EXTRA_DEVICE_ADMIN,
                    mAdminName);
            startActivity(intent);
        }
    }
    /**
     * 单击设置向导的第4个界面中的"设置完成"时所执行的方法，当执行该方法时，说明设
       置向导已经完成
     * @param view
     */
    public void next(View view){
        if(!cb_setup4_protect.isChecked()){//如果防盗保护没有开启，弹出一个对
                                            话框提示开启保护
            AlertDialog.Builder builder = new Builder(this);
            builder.setTitle("温馨提示");
            builder.setMessage("手机防盗极大地保护了你的手机安全，强烈建议开启!");
            builder.setPositiveButton("开启", new OnClickListener() {
                
                public void onClick(DialogInterface dialog, int which) {
                    //将防盗保护开启
                    cb_setup4_protect.setChecked(true);
        //设置向导已经完成，在用户下次进入时判断，值为true，说明已经进行过设置向导
                    Editor editor = sp.edit();
                    editor.putBoolean("issetup", true);
                    editor.commit();
                }
```

```
            });
            builder.setNegativeButton("取消", new OnClickListener() {

                public void onClick(DialogInterface dialog, int which) {
                    finish();
//设置向导已经完成，在用户下次进入时判断，值为true，说明已经进行过设置向导
                    Editor editor = sp.edit();
                    editor.putBoolean("issetup", true);
                    editor.commit();
                }
            });
            builder.create().show();
            return ;
        }
//设置向导已经完成，在用户下次进入时判断，值为true，说明已经进行过设置向导
        Editor editor = sp.edit();
        editor.putBoolean("issetup", true);
        editor.commit();
        Intent intent = new Intent(this,LostProtectedActivity.class);
        startActivity(intent);
        finish();
//自定义一个平移的动画效果。参数一：界面进入时的动画效果；参数二：界面出去时
    的动画效果
        overridePendingTransition(R.anim.tran_in, R.anim.tran_out);
    }
    public void pre(View view){
        Intent intent = new Intent(this,Setup3Activity.class);
        startActivity(intent);
        finish();
//自定义一个透明度变化的动画效果。参数一：界面进入时的动画效果；参数二：界面
    出去时的动画效果
        overridePendingTransition(R.anim.alpha_in, R.anim.alpha_out);
    }
}
```

代码解析：

ComponentName mAdminName = new ComponentName(this, MyAdmin.class)

MyAdmin 是一个用于获取超级管理员权限要被激活的组件。该组件一旦被激活，当前的应用程序就获取到了超级管理员的权限，就可以完成对手机的锁频、重置为出厂设置的操作。其实，这里的 MyAdmin 是一个特殊的广播接收者，MyAdmin 需要继承 DeviceAdminReceiver。

MyAdmin.java 代码如下：

```
package com.guoshisp.mobilesafe.receiver;
import android.app.admin.DeviceAdminReceiver;
```

```java
//DeviceAdminReceiver继承了广播接收者，激活设备超级管理员的权限
public class MyAdmin extends DeviceAdminReceiver {

}
```

然后，需要在清单文件中为该组件进行相应的配置：

```xml
<receiver android:name=".receiver.MyAdmin" >
        <meta-data
            android:name="android.app.device_admin"
            android:resource="@xml/my_admin" />
        <intent-filter>
            <action android:name="android.app.action.DEVICE_ADMIN_
                                   ENABLED" />
        </intent-filter>
</receiver>
```

代码解析：

（1）android:resource="@xml/my_admin"

my_admin.xml 中用于配置超级管理权限的信息，该文件位于 res\xml 目录下，该文件中的配置信息如下：

```xml
<?xml version="1.0" encoding="utf-8"?>
<device-admin xmlns:android="http://schemas.android.com/apk/res/android">
  <uses-policies>
    <limit-password />
    <watch-login />
    <reset-password />
    <force-lock />
    <wipe-data />
    <expire-password />
    <encrypted-storage />
    <disable-camera />
  </uses-policies>
    </device-admin>
```

（2）<action android:name="android.app.action.DEVICE_ADMIN_ENABLED" />

允许当前应用程序获取超级管理员的权限。

测试运行：当在设置向导的第 4 个界面激活 admin 时，效果图如图 2-24 所示，我们单击"激活"按钮即可激活当前应用程序的超级管理员权限。

向该设备发送信息"#*lockscreen*#"，锁屏的界面效果如图 2-25 所示（Android4.2 模拟器上首先会出现一个黑屏，这里是单击 Home 键之后的界面），当解锁时需要输入设置的密码 123 才可以解锁，解锁后如图 2-26 所示。

第 2 章 手机防盗模块的设计

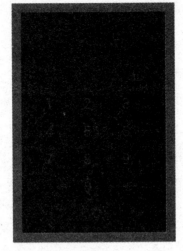

图 2-24　　　　　　　　　　图 2-25　　　　　　　　　　图 2-26

至此，手机防盗模块的业务逻辑全部实现完成。

需要掌握的知识点小结

在此，对前面使用到的重要知识点做以下小结：

（1）使用 Pull 解析器来解析服务端的 XML。

（2）连网时使用 httpurlconn，请求方式有 GET 和 POST。

（3）使用 PackageManager 获取程序的版本号信息。

（4）Intent（显式意图、隐式意图）。

（5）子线程和主线程通信时使用 Handler+Message。

（6）判断代码执行的时间（完成动画的播放）。

（7）对 Sdcard 文件的操作。

（8）自定义对话框。

（9）使用 GridView 显示九宫格，使用 selector 定义颜色的状态变化，使用 shape 来自定义图形。

（10）大量使用 SharedPreference 来持久化一些常用的数据。

（11）使用广播接收者来拦截外拨电话与快速进入手机防盗模块。

（12）MD5 加密算法。

（13）在设置向导中使用样式来简化布局文件的书写。

（14）使用 overridePendingTransition() 实现 Activity 切换时的动画效果。

（15）选择联系人时使用的三张表的关系。

（16）获取联系人时，通过内容提供者（uri，数据的映射关系）来获取联系人信息。

（17）使用 startactivityForResult() 让 Activity 返回数据。

（18）deviceadmin（设备的超级管理员）。

（19）GPS 位置的提供者。

（20）AlertDialog（注意两种上、下文的区别）。

第3章 高级工具模块的设计

3.1 号码归属地数据库的优化和复制

功能概述:在"高级工具"模块中,我们需要实现三个功能(号码归属地查询、常用号码查询、程序锁)。

号码归属地查询:有两种方式可以实现该功能,一种是通过 Webservice 来实现,另一种是通过本地数据库的查询。如果使用 Webservice,手机需要处在一个网络良好的状态才可以正常使用;如果从本地数据库中查询,由于未经压缩的完整的号码归属地数据库的体积大约有 20 MB,用户对应用的大小难以接受。但市场上主流的应用软件查询归属地的功能大都采用第二种方式。因为,这些归属地信息有很多的重复,我们可以将重复的数据抽取出来,同时通过外键引用的方式来查询数据库。这样,数据库的体积将会大大减少(压缩工具也正是采用提取公用的数据的做法来实现的)。

常用号码查询:将常用的号码展现出来,如订餐电话、公共服务、快递服务、机票酒店、银行证券等。当我们单击对应的号码时就可以实现拨号通话。

程序锁:列出当前手机中的所有应用程序,单击每个应用程序的 item 可以实现对应用程序的加锁、解锁,当应用程序被加锁后,且"设置中心"中的程序锁功能开启,那么,当打开该应用时,需要正确输入密码才能进入。

高级工具模块的 UI 预览图如图 3-1 所示。

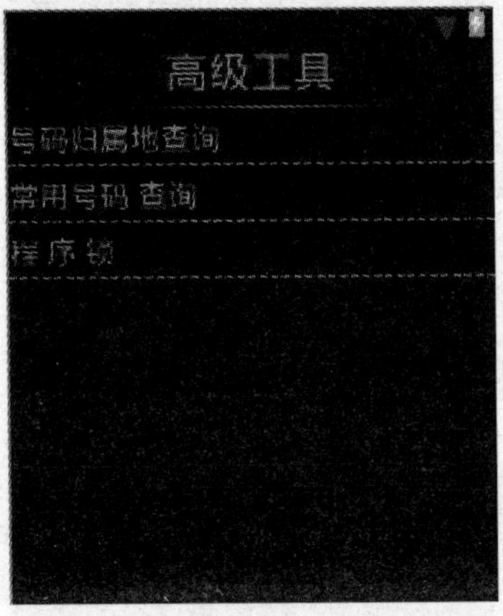

图 3-1

数据库的来源：可以到淘宝商城上购买该数据库。购买成功后，得到的是一个 Excel 表，我们可以通过 Oracle 将 Excel 文件中的数据生成 SQL 语句，通过 SQL 语句可以创建出 Android 下的.db 文件（address2.db 文件的大小是 16.4 MB）。这里使用的归属地信息的数据库没有包含 186 字段。使用 SQLite Expert 工具打开该数据库，如图 3-2 所示。在图 3-2 中我们可以看到有两张表：android_metadata 和 info，下面简单介绍这两张表。

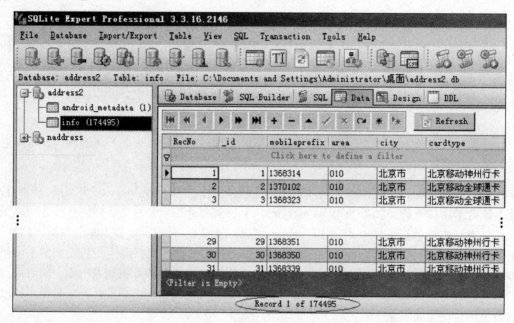

图 3-2

android_metadata 表：如果想在 Android 系统中使用数据库，数据库中必须存在一张 "android_metadata" 表，该表是为了以后做扩展使用的（例如，国际化）。

info：存储的是号码归属地的具体信息，由表名可以得出，号码归属地信息一共有 174495 条，如果包含 186 号段的话，大概有 25 万条数据。

通过对图 3-2 中的数据库表的观察可以发现：数据库中有很大一部分数据是重复的（字段 area、city、cardtype）。我们可以将这些重复的数据提取出来存放到另外一个表中，当查询时，通过外键引用即可实现。我们在 SQLite Expert 工具中的 SQL 下执行"select * from info group by area,city,cardtype"语句查看执行的结果集，而结果集中只有 2872 条数据，如图 3-3 所示。我们可以将该表单独创建出来，然后通过外键引用的方式来查询归属地信息。对 Android 下的数据库的优化做个小结：

➢ 提取冗余数据；
➢ 拆分冗余数据到一张新的表中；
➢ 利用外键引用实现查询。

使用以上方式对数据库进行优化，优化后的数据库为 naddress.db。优化后的数据库应当有三张表：android_metadata，numinfo 和 address_tb 表，其中，表 address_tb 是我们为提取的冗余数据而单独建立的一张表（通过表中的数据条数可以看出）。优化后的表 numinfo 如图 3-4

所示，表 address_tb 如图 3-5 所示。可以很明显地看出：经过优化后的表 numinfo 精简了许多。而优化后的数据库的大小只有 3.21 MB，不到之前的五分之一！

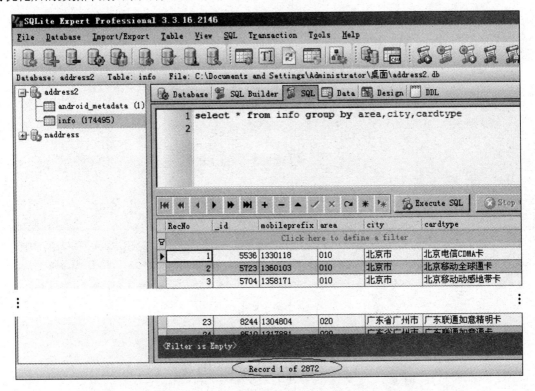

图 3-3

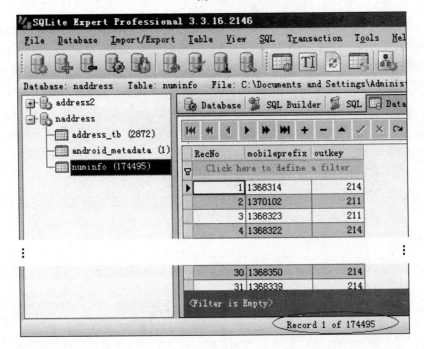

图 3-4

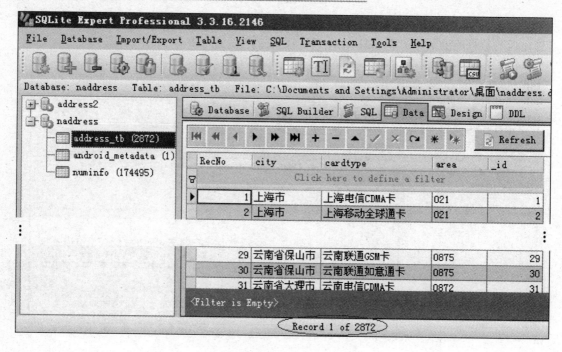

图 3-5

3.2 号码归属地查询

在 Android 中打开一个已存在的数据库的方式是 SQLiteDatabase.openDatabase (path, factory, flags)。使用数据库的前提是数据库文件必须已经被复制到系统中。如果我们在代码中创建出一个数据库，该数据库会被复制进手机的系统中。所以可以通过上面的语句直接打开使用。但是，我们这里的数据库是复制过来的。如果我们将数据库文件直接复制到工程的 assets 或 res 目录下，最终的数据库文件只会存在于 apk 中，而不会被释放到手机的系统中。我们需要将其读取到手机系统中才可以正常使用（处理方式是将该数据库存放在 assets 目录下，然后将其读取到手机的 data\data\应用程序包名\files 目录下，这样，数据库就复制到手机的系统中了）。以下是在 Android 中引用素材（文件）的三种方式：

（1）assets——资产目录（该目录下的素材资源不会在 R 文件中产生任何引用，会被直接打包到 apk 中）。获取该目录下的素材资源的方式是 nputStream is = getContext().getAssets(). open ("naddress.db")。

（2）res 下面的 raw 的目录（资源文件系统的 RescoureManager 会对资源产生引用，在 R 文件中可以查看到对应的引用值）。获取该目录下的素材资源的方式是 InputStream is = getContext().getResources().openRawResource(R.raw.naddress)。

（3）采用类加载器使用一个资源文件（javase 中的方式，将素材存放与 src 目录下。该方式在 sdk 开发中会被使用到）。获取该目录下的素材资源的方式是 InputStream is = getClass().

getClassLoader().getResourceAsStream("naddress.db")。

号码归属地查询的思路：当单击"号码归属地查询"条目时，首先会判断号码归属地信息的数据库是否已经被复制到系统中，如果系统中存在该文件，则进入号码归属地查询的界面，否则，执行复制动作，当复制成功后直接进入号码归属地查询界面。当进入查询界面后，在号码输入框中输入号码并单击"确定"按钮后，此时会根据输入的号码执行一个查询本地数据库的操作。

高级工具模块对应的 Activity 为 AtoolsActivity，AtoolsActivity.java 代码如下：

```java
package com.guoshisp.mobilesafe;
import java.io.File;
import android.app.Activity;
import android.app.ProgressDialog;
import android.content.Intent;
import android.os.Bundle;
import android.os.Handler;
import android.os.Message;
import android.view.View;
import android.view.View.OnClickListener;
import android.widget.TextView;
import android.widget.Toast;
import com.guoshisp.mobilesafe.utils.AssetCopyUtil;
public class AtoolsActivity extends Activity implements OnClickListener {
    protected static final int COPY_SUCCESS = 30;
    protected static final int COPY_FAILED = 31;
    private TextView tv_atools_address_query;//当单击该条目时，要执行复制号码
                                              归属地信息的数据库文件
    private ProgressDialog pd;//复制数据库时要显示的进度条
    //复制数据库是一个相对耗时的操作，复制完成后，给主线程发送消息
    private Handler handler = new Handler() {
        public void handleMessage(android.os.Message msg) {
            //无论复制是否成功，都需要关闭进度显示条
            pd.dismiss();
            switch (msg.what) {
            case COPY_SUCCESS:
                //复制数据库成功后，进入号码归属地查询的界面
                loadQueryUI();
                break;
            case COPY_FAILED:
                Toast.makeText(getApplicationContext(), "复制数据失败",0).
                          show();
                break;
            }
        };
    };
    @Override
```

```java
protected void onCreate(Bundle savedInstanceState) {
    super.onCreate(savedInstanceState);
    setContentView(R.layout.atools);//高级工具对应的界面
    tv_atools_address_query = (TextView) findViewById(R.id.tv_atools_
                    address_query);
    tv_atools_address_query.setOnClickListener(this);
    pd = new ProgressDialog(this);
    //设置进度条显示的风格
    pd.setProgressStyle(ProgressDialog.STYLE_HORIZONTAL);
}
//处理高级工具中的"号码归属地查询"的单击事件
public void onClick(View v) {
    switch (v.getId()) {
    case R.id.tv_atools_address_query:
        //创建出数据库要复制到的系统文件：data\data\包名\files\address.db
        final File file = new File(getFilesDir(), "address.db");
        //判断数据库是否存在，如果存在，则直接进入号码归属地的查询界面，否则，执
          行复制动作
        if (file.exists() && file.length() > 0) {
            //数据库文件复制成功，进入查询号码归属地界面
            loadQueryUI();
        } else {
            //数据库的复制，开始复制时需要开始显示进度条
            pd.show();
            //复制数据库也是一个相对耗时的操作，在子线程中执行该操作
            new Thread() {
                public void run() {
                    AssetCopyUtil asu = new AssetCopyUtil(
                            getApplicationContext());
                    //返回复制成功与否的结果
                    boolean result = asu.copyFile("naddress.db", file,pd);
                    if (result) {//复制成功
                        Message msg = Message.obtain();
                        msg.what = COPY_SUCCESS;
                        handler.sendMessage(msg);
                    } else {//复制失败
                        Message msg = Message.obtain();
                        msg.what = COPY_FAILED;
                        handler.sendMessage(msg);
                    }
                }
            }.start();
        }
        break;
    }
}
```

```java
/**
 * 进入到号码归属地查询界面
 */
private void loadQueryUI() {
    Intent intent = new Intent(this, NumberQueryActivity.class);
    startActivity(intent);
}
}
```

布局文件存放在 layout 文件夹下，atools.xml 文件如下：

```xml
<?xml version="1.0" encoding="utf-8"?>
<LinearLayout xmlns:android="http://schemas.android.com/apk/res/android"
    android:layout_width="fill_parent"
    android:layout_height="fill_parent"
    android:orientation="vertical" >
    <TextView
        android:layout_width="wrap_content"
        android:layout_height="wrap_content"
        android:layout_gravity="center_horizontal"
        android:text="高级工具"
        android:textColor="#66ff00"
        android:textSize="28sp" />
    <View
        android:layout_width="fill_parent"
        android:layout_height="1dip"
        android:layout_marginTop="5dip"
        android:background="@drawable/devide_line" />
    <TextView
         android:layout_marginTop="8dip"
        android:id="@+id/tv_atools_address_query"
        style="@style/text_content_style"
        android:text="号码归属地查询" />
    <View
        android:layout_width="fill_parent"
        android:layout_height="1dip"
        android:layout_marginTop="4dip"
        android:background="@drawable/listview_devider" >
    </View>
</LinearLayout>
```

当第一次进入时，需要复制数据库。复制数据库的对象 AssetCopyUtil 存放在包 "com.guoshisp.mobilesafe.utils" 下，AssetCopyUtil.java 对应的代码为：

```java
package com.guoshisp.mobilesafe.utils;
import java.io.File;
import java.io.FileOutputStream;
import java.io.IOException;
```

```java
import java.io.InputStream;
import android.app.ProgressDialog;
import android.content.Context;
import android.content.res.AssetManager;
/**
 * 资产文件复制的工具类
 * @author
 *
 */
public class AssetCopyUtil {
    private Context context;
    public AssetCopyUtil(Context context) {
        this.context = context;
    }
    /**
     * 复制资产目录下的文件
     * @param srcfilename 源文件的名称
     * @param file 目标文件的对象
     * @param pd 进度条对话框
     * @return 是否复制成功
     */
    public boolean copyFile(String srcfilename,File file,ProgressDialog pd){
        try {
            //获取到资产目录的管理器,因为数据库存放在该目录下
            AssetManager am = context.getAssets();
            //打开资产目录下的资源文件,获取一个输入流对象
            InputStream is = am.open(srcfilename);
            //获取到该文件的字节数
            int max = is.available();
            //设置进度条显示的最大进度
            pd.setMax(max);
            //创建一个输出流文件,用于接收输入流
            FileOutputStream fos = new FileOutputStream(file);
            //创建一个缓存区
            byte[] buffer = new byte[1024];
            int len=0;
            //进度条的最开始的位置应该为 0
            int process = 0;
            while((len = is.read(buffer))!=-1){
                fos.write(buffer, 0, len);
                //让进度条不断地动态显示当前的复制进度
                process+=len;
                pd.setProgress(process);
            }
            //刷新缓冲区,关流
            fos.flush();
```

```
            fos.close();
            return true;
        } catch (Exception e) {
            e.printStackTrace();
            return false;
        }
    }
}
```

说明：在该工具代码中将异常"吃掉"的原因在于异常所在方法的返回值是一个 boolean 类型的值，可以根据返回的 boolean 值来判断复制数据库这个动作是否成功。

如果数据库已经存在于系统中，应该直接进入号码归属地查询的界面，号码归属地查询对应的 Activity 为 NumberQueryActivity，NumberQueryActivity.java 的业务代码如下：

```java
package com.guoshisp.mobilesafe;
import android.app.Activity;
import android.os.Bundle;
import android.text.TextUtils;
import android.view.View;
import android.view.animation.Animation;
import android.view.animation.AnimationUtils;
import android.widget.EditText;
import android.widget.TextView;
import android.widget.Toast;
import com.guoshisp.mobilesafe.db.dao.NumberAddressDao;
public class NumberQueryActivity extends Activity {
    private EditText et_number_query;//输入要查询的号码
    private TextView tv_number_address;//显示号码归属地位置
    @Override
    protected void onCreate(Bundle savedInstanceState) {
        setContentView(R.layout.number_query);
        super.onCreate(savedInstanceState);
        et_number_query = (EditText) findViewById(R.id.et_number_query);
        tv_number_address = (TextView) findViewById(R.id.tv_number_address);
    }
    /**
     * 单击"查询"时执行的监听方法
     *
     * @param view
     */
    public void query(View view) {
        //查询前，需要将号码前后的空格清空
        String number = et_number_query.getText().toString().trim();
        //判断要查询的号码是否为空
        if (TextUtils.isEmpty(number)) {
            Toast.makeText(this, "号码不能为空", 1).show();
```

```
            //使用动画工具来加载一个动画资源一个动画资源
            Animation shake = AnimationUtils.loadAnimation(this, R.anim.
                shake);
            //当号码输入框中没有输入号码而单击"查询"时播放一个动画，用来提示用户输
              入号码后才可以执行查询操作。
            et_number_query.startAnimation(shake);
            return;
        } else {//号码不为空时要返回归属地信息
            //返回查询到的归属地信息
            String address = NumberAddressDao.getAddress(number);
            //将归属地信息显示在屏幕上
            tv_number_address.setText(address);
        }
    }
}
```

布局文件存放在 layout 文件夹下，number_query.xml 文件如下：

```xml
<?xml version="1.0" encoding="utf-8"?>
<LinearLayout xmlns:android="http://schemas.android.com/apk/res/android"
    android:layout_width="fill_parent"
    android:layout_height="fill_parent"
    android:orientation="vertical" >
    <TextView
        android:layout_width="wrap_content"
        android:layout_height="wrap_content"
        android:layout_gravity="center_horizontal"
        android:text="号码归属地查询"
        android:textColor="#66ff00"
        android:textSize="28sp" />
    <View
        android:layout_width="fill_parent"
        android:layout_height="1dip"
        android:layout_marginTop="5dip"
        android:background="@drawable/devide_line" />
    <EditText
        android:inputType="phone"
        android:hint="请输入查询的电话号码"
        android:layout_marginTop="8dip"
        android:id="@+id/et_number_query"
        android:layout_width="match_parent"
        android:layout_height="wrap_content" >
    </EditText>
    <Button
        android:onClick="query"
        android:layout_width="wrap_content"
        android:layout_height="wrap_content"
```

```
        android:text="查询" />
    <TextView
        android:id="@+id/tv_number_address"
        android:layout_width="wrap_content"
        android:layout_height="wrap_content"
        android:text="归属地:"
        android:textAppearance="?android:attr/textAppearanceLarge" />
</LinearLayout>
```

当号码输入框中没有输入号码而单击"查询"按钮时播放一个动画,该动画资源位于 res\anim 目录下,shake.xml 文件内容如下:

```
<?xml version="1.0" encoding="utf-8"?>
<translate xmlns:android="http://schemas.android.com/apk/res/android"
    android:duration="1000"
    android:fromXDelta="0"
    android:interpolator="@anim/cycle_7"
    android:toXDelta="10" />
```

其中,android:interpolator="@anim/cycle_7"代表的是动画的具体表现形式,cycle_7.xml 位于 res\anim 目录下,cycle_7.xml 文件内容如下:

```
<?xml version="1.0" encoding="utf-8"?>
<cycleInterpolator xmlns:android="http://schemas.android.com/apk/res/
                                  android"
    android:cycles="7" />
```

说明:cycle_7.xml 文件所定义的是动画的具体表现形式,引用的是系统资源。其中 cycleInterpolator 代表的是圆环加速器,android:cycles="7"表示所作用的控件要抖动 7 次。所以, shake.xml 的动画资源效果应该是:在 1000 毫秒内,控件应当连续左右晃动 7 次,每次在 X 轴 方向上晃动的距离为 10 个像素。

代码解析:

String address = NumberAddressDao.getAddress(number)

当输入所要查询的号码后,需要进入号码归属地的数据库中来查询将该号码的归属地信息。执行该操作的对象 NumberAddressDao 存放在包 "com.guoshisp.mobilesafe.db.dao" 下, NumberAddressDao.java 中的代码如下:

```java
package com.guoshisp.mobilesafe.db.dao;
import android.database.Cursor;
import android.database.sqlite.SQLiteDatabase;
/**
 * 号码归属地查询——操作数据库
 * @author
 */
public class NumberAddressDao {
    /**
```

```java
 * 获取电话号码的归属地
 * @param number
 * @return
 */
public static String getAddress(String number) {
    //如果没有查询到号码的归属地，就返回当前的电话号码
    String address = number;
    //数据库在手机系统中的全路径
    String path = "/data/data/com.guoshisp.mobilesafe/files/address.db";
    //打开数据库。参数二：CursorFactory游标工厂，null 表示使用系统默认
    SQLiteDatabase db = SQLiteDatabase.openDatabase(path, null,
            SQLiteDatabase.OPEN_READONLY);
    //判断数据库是否被打开
    if (db.isOpen()) {
        //判断号码的类型
        if (number.matches("^1[3458]\\d{9}$")) {
            //返回一个查询的结果集
            Cursor cursor = db
                    .rawQuery(
                            "select city from address_tb where _id=
            (select outkey from numinfo where mobileprefix =?)",
                            //匹配手机的前 7 位
                            new String[] { number.substring(0, 7) });
    if (cursor.moveToFirst()) {
                address = cursor.getString(0);//获取第 0 列即可
            }
            cursor.close();
        } else {//其他号码（固定电话）
            Cursor cursor;
            switch (number.length()) {//号码的长度
            case 4:
                address = "模拟器";
                break;
            case 7://本地号码不显示区号
                address = "本地号码";
                break;
            case 8:
                address = "本地号码";
                break;
            case 10:
                //从查询返回结果中获取第一条数据（limit 表示只获取第一条数据）
                cursor = db
                        .rawQuery(
                                "select city from address_tb where area
                                 = ? limit 1",
                                new String[] { number.substring(0, 3) });
```

```java
            if (cursor.moveToFirst()) {
                address = cursor.getString(0);
            }
            cursor.close();
            break;
        case 12://4 位的区号+8 位的号码
            cursor = db
                    .rawQuery(
                            "select city from address_tb where area = ? limit 1",
                            new String[] { number.substring(0, 4) });
            if (cursor.moveToFirst()) {
                address = cursor.getString(0);
            }
            cursor.close();
            break;
        case 11://3 位的区号+8 位的号码,或者是 4 位的区号+7 位的号码
            cursor = db
                    .rawQuery(
                            "select city from address_tb where area = ? limit 1",
                            new String[] { number.substring(0, 3) });
            if (cursor.moveToFirst()) {
                address = cursor.getString(0);
            }
            cursor.close();
            cursor = db
                    .rawQuery(
                            "select city from address_tb where area = ? limit 1",
                            new String[] { number.substring(0, 4) });
            if (cursor.moveToFirst()) {
                address = cursor.getString(0);
            }
            cursor.close();
            break;
        }
    }
    db.close();
}
return address;
```

说明：

查询手机号码归属地的信息：首先从表 numinfo 中获取该号码的外键 outkey，得到 outkey

之后，再从表 addres_tb 中查询其对应的归属地。由 naddress.db 数据库中的表结构可以得出查询手机号码归属地的 SQL 语句。

获取外键：select outkey from numinfo where mobileprefix ='1351234'。该语句的执行结果为 2631。通过对外键的引用来查询归属地信息：select city from address_tb where _id='2631'。可将两句合并为：select city from address_tb where _id=(select outkey from numinfo where mobileprefix='1358888')。

查询固定电话的归属地信息：由于固定电话号码前面存在区号，所以，只需要在 address_tb 表中通过区号（area）来判断归属地（city）即可。由 naddress.db 数据库中的表结构可以得出查询手机号码归属地的 SQL 语句：select city from address_tb where area = '020' limit 1，其中，limit 1 表示从查询的结果中只获取第一列数据。

代码解析：

（1）number.matches("^1[3458]\\d{9}$")

这里通过采用正则表达式来判断手机号码的具体类型。

- "^"：匹配输入字符串的开始位置。如果设置了 RegExp 对象的 Multiline 属性，"^" 也匹配'\n'或'\r'之后的位置。"^1" 则表示开始的字符应该是 1。

- "[xyz]"：字符集合。匹配所包含的任意一个字符。例如，'[abc]'可以匹配"plain"中的'a'。[3458]表示第二个字符可以是 3，4，5，8 中的任一个都行。

- "\"：将下一个字符标记为一个特殊字符、或一个原义字符、或一个后向引用、或一个八进制转义符。例如，'n'匹配字符"n"。'\n'匹配一个换行符。序列'\\'匹配"\"而"\("则匹配"("。

- "\d"：匹配一个数字字符，等价于[0~9]。

- "{n}"：n 是一个非负整数。匹配确定的 n 次。例如，'o{2}'不能匹配"Bob"中的'o'，但是能匹配"food"中的两个 o。\d{9}表示 0~9 之间的数字出现了 9 次。

- "$"：匹配输入字符串的结束位置。如果设置了 RegExp 对象的 Multiline 属性，$也匹配'\n'或'\r'之前的位置。

（2）Cursor cursor = db
    .rawQuery(
        "select city from address_tb where _id=(select outkey from numinfo where mobileprefix =?)",
        new String[] { number.substring(0, 7) })

通过对 SQL 语句的查询返回一个 Cursor 结果集。"mobileprefix =?" 是将 mobileprefix 用一个占位符先表示一下，在 new String[] { number.substring(0, 7) }中给出占位符的具体值。number.substring(0, 7)是截取 number 中的前 7 位字符。

另外，我们需要在清单文件中为相应的组件配置组件信息。还需要在主界面为"高级工具"

item 设置单击事件的处理。MainActivity.java 的业务代码如下：

```java
package com.guoshisp.mobilesafe;
import android.app.Activity;
import android.content.Intent;
import android.os.Bundle;
import android.view.View;
import android.widget.AdapterView;
import android.widget.AdapterView.OnItemClickListener;
import android.widget.GridView;
import com.guoshisp.mobilesafe.adapter.MainAdapter;
public class MainActivity extends Activity {
    //显示主界面中九大模块的GridView
    private GridView gv_main;
    @Override
    protected void onCreate(Bundle savedInstanceState) {
        super.onCreate(savedInstanceState);
        setContentView(R.layout.main);
        gv_main = (GridView) findViewById(R.id.gv_main);
        //为gv_main对象设置一个适配器，该适配器的作用是为每个item填充对应的数据
        gv_main.setAdapter(new MainAdapter(this));
        //为GridView对象中的item设置单击时的监听事件
        gv_main.setOnItemClickListener(new OnItemClickListener() {
            //参数一：item的父控件，也就是GridView；参数二：当前单击的item；参
            //  数三：当前单击的item在GridView中的位置
            //参数四：id的值为单击GridView的哪一项对应的数值，单击GridView
            //  第9项，那id就等于8
            public void onItemClick(AdapterView<?> parent, View view,
                    int position, long id) {
                switch (position) {
                case 0: //手机防盗
                    //跳转到"手机防盗"对应的Activity界面
                    Intent lostprotectedIntent = new Intent(MainActivity.
                            this,LostProtectedActivity.class);
                    startActivity(lostprotectedIntent);
                    break;
                case 7://高级工具
                    Intent atoolsIntent = new Intent(MainActivity.this,
                            AtoolsActivity.class);
                    startActivity(atoolsIntent);
                    break;
                case 8://设置中心
                    //跳转到"设置中心"对应的Activity界面
                    Intent settingIntent = new Intent(MainActivity.this,
                            SettingCenterActivity.class);
                    startActivity(settingIntent);
                    break;
```

```
            }
          }
        });
      }
    }
```

测试运行：第一次进入"高级工具"中，单击"号码归属地查询"条目时的界面效果如图 3-6 所示。当数据库被复制到手机系统中进入到"号码归属地查询"界面，未输入号码单击"查询"按钮时的界面如图 3-7 所示。输入"13888888888"单击查询后（该号码归属地数据库是经典版的，没有包括所有的号码），界面效果如图 3-8 所示。

图 3-6

图 3-7

图 3-8

3.3 显示来电与外拨电话的号码归属地

显示来电归属地的需求：当一个号码呼叫进来时，需要在屏幕上以一个吐司将归属地信息展示出来。在电话未接通前，吐司一直显示。在设置中心中设置来电归属地信息显示是否开启、来电归属地显示风格、来电归属地信息显示的位置。

问题：我们知道，Android 系统提供的吐司的显示时长较短，不能够满足在未接通电话一直显示归属地信息的需求。

解决方案：通过自定义吐司显示来电归属地信息。

来电归属地信息显示实现的步骤：

（1）需要知道电话何时到来。可以在后台服务中为其注册一个监听器，来监听电话的状态。

（2）知道这个电话号码的 number。可以在监听器里面获取电话。

第3章 高级工具模块的设计

（3）把归属地查询出来。可以直接使用前面已经写好的查询号码归属地信息的工具代码。

（4）将归属地显示到界面上。可以采用自定义的吐司来展示归属地信息。

我们将外拨电话时显示归属地信息的服务 ShowCallLocationService 放在包"com.guoshisp.mobilesafe.service"下，并在清单文件中为其配置对应的组件信息。

ShowCallLocationService.java 的业务代码如下：

```java
package com.guoshisp.mobilesafe.service;
import android.app.Service;
import android.content.Intent;
import android.graphics.PixelFormat;
import android.os.IBinder;
import android.telephony.PhoneStateListener;
import android.telephony.TelephonyManager;
import android.view.View;
import android.view.WindowManager;
import android.view.WindowManager.LayoutParams;
import android.widget.LinearLayout;
import android.widget.TextView;
import android.widget.Toast;
import com.guoshisp.mobilesafe.R;
import com.guoshisp.mobilesafe.db.dao.NumberAddressDao;
//在后台监听电话呼入的状态
public class ShowCallLocationService extends Service {
    private TelephonyManager tm;//电话管理器
    private MyPhoneListener listener;//电话状态改变的监听器
    private WindowManager windowManager;//窗体管理器
    @Override
    public IBinder onBind(Intent intent) {
        return null;
    }
    /**
     * 当服务第一次被创建时调用
     */
    @Override
    public void onCreate() {
        super.onCreate();
        //注册系统的电话状态改变的监听器
        listener = new MyPhoneListener();
        //获取系统的电话管理器
        tm = (TelephonyManager) getSystemService(TELEPHONY_SERVICE);
        //为电话设置一个监听。参数一：监听器；参数二：要监听的电话改变类型（这里监听
            的是通话状态）
        tm.listen(listener, PhoneStateListener.LISTEN_CALL_STATE);
        windowManager = (WindowManager) getSystemService(WINDOW_SERVICE);
    }
```

```java
private class MyPhoneListener extends PhoneStateListener {
    private View view;
    //参数一：手机的状态；参数二：呼叫进来的手机号码
    @Override
    public void onCallStateChanged(int state, String incomingNumber) {
        switch (state) {
            case TelephonyManager.CALL_STATE_RINGING://手机铃声正在响
                //获取呼叫进来号码的地址（查询之前的号码归属地数据库）
                String address = NumberAddressDao.getAddress(incomingNumber);
                //使用系统的吐司来显示归属地信息，但显示的时间较短
                //Toast.makeText(getApplicationContext(), "归属地:"+address, 1).show();
                //通过布局填充器将一个显示号码归属地的布局转成View,该View是一个吐司
                view = View.inflate(getApplicationContext(), R.layout.
                    show_address,null);
                //获取到显示号码归属地布局的根布局LinearLayout
                LinearLayout ll = (LinearLayout) view.findViewById(R.id.
                        ll_show_address);
                //查找到View中用于显示归属地的TextView
                TextView tv = (TextView) view.findViewById(R.id.tv_show_
                        address);
                //将归属地信息设置到TextView
                tv.setText(address);
                //获取到与窗体相关的布局的参数（这里用于设置窗体上显示来电归属地的吐司的参数信息）
                final WindowManager.LayoutParams params = new LayoutParams();
                //设置窗体布局View的高度
                params.height = WindowManager.LayoutParams.WRAP_CONTENT;
                //设置窗体布局View的宽度
                params.width = WindowManager.LayoutParams.WRAP_CONTENT;
                //窗体View不可以获取焦点、不可以被触摸、保持在屏幕上
                params.flags = WindowManager.LayoutParams.FLAG_NOT_FOCUSABLE
                    | WindowManager.LayoutParams.FLAG_NOT_TOUCHABLE
                    | WindowManager.LayoutParams.FLAG_KEEP_SCREEN_ON;
                //显示在窗体上的style为半透明
                params.format = PixelFormat.TRANSLUCENT;
                //窗体View的类型为吐司
                params.type = WindowManager.LayoutParams.TYPE_TOAST;
                //将吐司挂载在窗体上
                windowManager.addView(view, params);
                break;
            case TelephonyManager.CALL_STATE_IDLE: //手机的空闲状态
                break;
            case TelephonyManager.CALL_STATE_OFFHOOK://手机接通通话时的状态
                break;
        }
        super.onCallStateChanged(state, incomingNumber);
    }
}
```

```java
        }
    /**
     * 取消电话状态的监听
     */
    @Override
    public void onDestroy() {
        super.onDestroy();
        tm.listen(listener, PhoneStateListener.LISTEN_NONE);
        listener = null;
    }
}
```

自定吐司的思路：首先在 layout 文件夹下创建一个普通的布局文件 show_address.xml，通过布局填充器将其转换为一个 View 对象，然后获取窗体布局参数对象，为该 View 对象设置一系列的参数，最后为窗体添加一个 View 对象。在添加 View 时，将 View 和窗体布局参数对象传递进去即可显示这个 View 吐司。show_address.xml 布局文件如下：

```xml
<?xml version="1.0" encoding="utf-8"?>
<LinearLayout xmlns:android="http://schemas.android.com/apk/res/android"
    android:layout_width="wrap_content"
    android:layout_height="wrap_content"
    android:gravity="center_vertical"
    android:id="@+id/ll_show_address"
    android:orientation="horizontal"
    android:background="@drawable/call_locate_white"
    >
    <ImageView
        android:layout_width="wrap_content"
        android:layout_height="wrap_content"
        android:src="@drawable/notification" />
    <TextView
        android:id="@+id/tv_show_address"
        android:layout_width="wrap_content"
        android:layout_height="wrap_content"
        android:textSize="20sp"
        android:textColor="@android:color/white"
        android:text="归属地" />
</LinearLayout>
```

以上实现了一个吐司的 View 布局。接下来，需要在设置中心界面中完成来电归属地信息显示是否开启的设置。

SettingCenterActivity.java 业务代码如下：

```java
package com.guoshisp.mobilesafe;
import android.app.Activity;
import android.content.Intent;
import android.content.SharedPreferences;
```

```java
import android.content.SharedPreferences.Editor;
import android.graphics.Color;
import android.os.Bundle;
import android.view.View;
import android.view.View.OnClickListener;
import android.widget.CheckBox;
import android.widget.CompoundButton;
import android.widget.CompoundButton.OnCheckedChangeListener;
import android.widget.RelativeLayout;
import android.widget.TextView;
import com.guoshisp.mobilesafe.service.ShowCallLocationService;
public class SettingCenterActivity extends Activity implements
    OnClickListener {
    //用于存储自动更新是否开启的boolean值
    private SharedPreferences sp;
    //自动更新的是否开启对应的TextView控件的显示文字
    private TextView tv_setting_autoupdate_status;
    //显示自动更新是否开启的勾选框
    private CheckBox cb_setting_autoupdate;
    //归属地显示控件的声明
    private TextView tv_setting_show_location_status;//显示来显归属地是否开
                                                     启的状态
    private CheckBox cb_setting_show_location;//是否开启来电归属地的Checkbox
    private RelativeLayout rl_setting_show_location;//"来电归属地是否开启"
                                                     控件的父控件
    private Intent showLocationIntent;//开启来电归属地信息显示的意图
    @Override
    protected void onCreate(Bundle savedInstanceState) {
        setContentView(R.layout.setting_center);
        super.onCreate(savedInstanceState);
        //获取Sdcard下的config.xml文件,如果该文件不存在,那么将会自动创建该文
          件,文件的操作类型为私有类型
        sp = getSharedPreferences("config", MODE_PRIVATE);
        //标记自动更新状态是否开启对应的Checkbox控件
        cb_setting_autoupdate = (CheckBox) findViewById(R.id.cb_setting_
                             autoupdate);
        //显示当前自动更新是否开启对应的TextView控件
        tv_setting_autoupdate_status = (TextView) findViewById(R.id.tv_
                             setting_autoupdate_status);
        //初始化自动更新的ui,默认状态下是开启的
        boolean autoupdate = sp.getBoolean("autoupdate", true);
        if (autoupdate) {
            tv_setting_autoupdate_status.setText("自动更新已经开启");
            //因为autoupdate变量为true,则表示自动更新开启,所以,Checkbox的
              状态应该是勾选状态的,即为true
            cb_setting_autoupdate.setChecked(true);
```

```java
        } else {
            tv_setting_autoupdate_status.setText("自动更新已经关闭");
            //因为autoupdate变量为false,则表示自动更新未开启,所以,Checkbox
               的状态应该是未勾选状态的,即为false
            cb_setting_autoupdate.setChecked(false);
        }
        /**
         * 当Checkbox的状态发生改变时执行以下代码
         */
        cb_setting_autoupdate
                .setOnCheckedChangeListener(new OnCheckedChangeListener() {
            //参数一:当前的Checkbox;参数二:当前的Checkbox是否处于勾选状态
                    public void onCheckedChanged(CompoundButton buttonView,
                            boolean isChecked) {
            //获取编辑器
                        Editor editor = sp.edit();
            //持久化存储当前Checkbox的状态,当下次进入时,依然可以保存当前设置的状态
                        editor.putBoolean("autoupdate", isChecked);
                        //将数据真正提交到sp里面
                        editor.commit();
                        if (isChecked) {//Checkbox处于选中效果
            //当Checkbox处于勾选状态时,表示自动更新已经开启,同时修改字体颜色
                            tv_setting_autoupdate_status
                                    .setTextColor(Color.WHITE);
                            tv_setting_autoupdate_status.setText("自动更
                                新已经开启");
                        } else {//Checkbox处于未勾选状态
            //当Checkbox未处于勾选状态时,表示自动更新已经开启,同时修改字体颜色
                            tv_setting_autoupdate_status
                                    .setTextColor(Color.RED);
                            tv_setting_autoupdate_status.setText("自动更
                                新已经关闭");
                        }
                    }
                });
        //显示归属地信息的ui初始化
        tv_setting_show_location_status = (TextView) findViewById(R.id. tv_
                                    setting_show_location_status);
        cb_setting_show_location = (CheckBox) findViewById(R.id.cb_setting_
                            show_location);
        rl_setting_show_location = (RelativeLayout) findViewById(R.id.
                            rl_setting_show_location);
        showLocationIntent = new Intent(this, ShowCallLocationService.class);
        rl_setting_show_location.setOnClickListener(this);
    }
    //响应单击事件
```

```java
@Override
public void onClick(View v) {
    //TODO Auto-generated method stub
    switch (v.getId()) {
    case R.id.rl_setting_show_location://来电归属地是否开启
        if (cb_setting_show_location.isChecked()) {
            tv_setting_show_location_status.setText("来电归属地显示没
                有开启");
            stopService(showLocationIntent);
            cb_setting_show_location.setChecked(false);
        } else {
            tv_setting_show_location_status.setText("来电归属地显示已
                经开启");
            startService(showLocationIntent);
            cb_setting_show_location.setChecked(true);
        }
        break;
    }
}
```

其对应的 setting_center.xml 布局文件如下（该布局文件已经实现归属地信息的显示风格和显示位置的布局）：

```xml
<?xml version="1.0" encoding="utf-8"?>
<LinearLayout xmlns:android="http://schemas.android.com/apk/res/android"
    android:layout_width="fill_parent"
    android:layout_height="fill_parent"
    android:orientation="vertical" >
<TextView
    android:layout_width="wrap_content"
    android:layout_height="wrap_content"
    android:layout_gravity="center_horizontal"
    android:text="设置中心"
    android:textColor="#66ff00"
    android:textSize="28sp" />
<View
    android:layout_width="fill_parent"
    android:layout_height="1dip"
    android:layout_marginTop="5dip"
    android:background="@drawable/devide_line" />
<RelativeLayout
    android:layout_width="fill_parent"
    android:layout_height="wrap_content"
    android:layout_marginTop="8dip" >
    <TextView
        android:id="@+id/tv_setting_autoupdate_text"
```

```xml
            android:layout_width="wrap_content"
            android:layout_height="wrap_content"
            android:text="自动更新设置"
            android:textSize="26sp"
            android:textStyle="bold" >
        </TextView>
        <TextView
            android:id="@+id/tv_setting_autoupdate_status"
            android:layout_width="wrap_content"
            android:layout_height="wrap_content"
            android:layout_below="@id/tv_setting_autoupdate_text"
            android:text="自动更新没有开启"
            android:textColor="#ffffff"
            android:textSize="14sp" />
        <CheckBox
            android:id="@+id/cb_setting_autoupdate"
            android:layout_width="wrap_content"
            android:layout_height="wrap_content"
            android:layout_alignParentRight="true"
            android:checked="false" />
</RelativeLayout>
<View
    android:layout_width="fill_parent"
    android:layout_height="1dip"
    android:layout_marginTop="4dip"
    android:background="@drawable/listview_devider" >
</View>
<RelativeLayout
    android:id="@+id/rl_setting_show_location"
    android:layout_width="fill_parent"
    android:layout_height="wrap_content"
    android:layout_marginTop="8dip" >
    <TextView
        android:id="@+id/tv_setting_show_location"
        android:layout_width="wrap_content"
        android:layout_height="wrap_content"
        android:text="来电归属地设置"
        android:textSize="26sp"
        android:textStyle="bold" >
    </TextView>
    <TextView
        android:id="@+id/tv_setting_show_location_status"
        android:layout_width="wrap_content"
        android:layout_height="wrap_content"
        android:layout_below="@id/tv_setting_show_location"
        android:text="来电归属地显示没有开启"
```

```xml
            android:textColor="#ffffff"
            android:textSize="14sp" />
        <CheckBox
            android:id="@+id/cb_setting_show_location"
            android:layout_width="wrap_content"
            android:layout_height="wrap_content"
            android:layout_alignParentRight="true"
            android:checked="false"
            android:clickable="false"
            android:focusable="false" />
    </RelativeLayout>
    <View
        android:layout_width="fill_parent"
        android:layout_height="1dip"
        android:layout_marginTop="4dip"
        android:background="@drawable/listview_devider" >
    </View>
    <RelativeLayout
        android:id="@+id/rl_setting_change_bg"
        android:layout_width="fill_parent"
        android:layout_height="wrap_content"
        android:layout_marginTop="8dip" >
        <TextView
            android:id="@+id/tv_setting_change_bg"
            android:layout_width="wrap_content"
            android:layout_height="wrap_content"
            android:text="来电归属地风格设置"
            android:textSize="26sp"
            android:textStyle="bold" >
        </TextView>
        <TextView
            android:id="@+id/tv_setting_show_bg"
            android:layout_width="wrap_content"
            android:layout_height="wrap_content"
            android:layout_below="@id/tv_setting_change_bg"
            android:text="半透明"
            android:textColor="#ffffff"
            android:textSize="14sp" />
        <ImageView
            android:layout_width="wrap_content"
            android:layout_height="wrap_content"
            android:layout_alignParentRight="true"
            android:layout_centerVertical="true"
            android:clickable="false"
            android:focusable="false"
            android:src="@drawable/jiantou1" />
```

```xml
    </RelativeLayout>
    <View
        android:layout_width="fill_parent"
        android:layout_height="1dip"
        android:layout_marginTop="4dip"
        android:background="@drawable/listview_devider" >
    </View>
    <RelativeLayout
        android:id="@+id/rl_setting_change_location"
        android:layout_width="fill_parent"
        android:layout_height="wrap_content"
        android:layout_marginTop="8dip" >
        <TextView
            android:id="@+id/tv_setting_change_location"
            android:layout_width="wrap_content"
            android:layout_height="wrap_content"
            android:text="归属地提示框的位置"
            android:textSize="26sp"
            android:textStyle="bold" >
        </TextView>
        <TextView
            android:layout_width="wrap_content"
            android:layout_height="wrap_content"
            android:layout_below="@id/tv_setting_change_location"
            android:text="更改归属地提示框位置"
            android:textColor="#ffffff"
            android:textSize="14sp" />
        <ImageView
            android:layout_width="wrap_content"
            android:layout_height="wrap_content"
            android:layout_alignParentRight="true"
            android:layout_centerVertical="true"
            android:clickable="false"
            android:focusable="false"
            android:src="@drawable/jiantou1" />
    </RelativeLayout>
    <View
        android:layout_width="fill_parent"
        android:layout_height="1dip"
        android:layout_marginTop="4dip"
        android:background="@drawable/listview_devider" >
    </View>
</LinearLayout>
```

测试运行：首先在设置中心中开启"来电归属地设置"，然后进入 Eclipse 的 Emulator Control 中模拟一个呼入的手机号码 13888888888，界面效果图如图 3-9 所示，当挂断电话并按下 Home

键后，界面效果图如图 3-10 所示。

图 3-9   图 3-10

在图 3-10 中我们可以发现：显示来电归属地信息的吐司并没有随着电话的挂断而消失，如果我们再次呼入一个电话，那么将会出现吐司重叠的情况。原因在于：我们在窗体上挂载一个 View 时开启了窗体服务，而窗体服务是一个全局的系统服务，该服务开启后会在后台运行。一般情况下，在窗体上一旦挂载一个 View 并显示后，该 View 并不会自动消失。

解决方案：需要通过代码将该 View 从窗体上移除。这一步可以在手机处于空闲或通话状态时来实现，代码如下：

```
case TelephonyManager.CALL_STATE_IDLE: //手机的空闲状态
        if(view!=null){
            //将窗体上的吐司移除
            windowManager.removeView(view);
            view = null;
        }
        break;
case TelephonyManager.CALL_STATE_OFFHOOK://手机接通通话时的状态
        if(view!=null){
            //将窗体上的吐司移除
            windowManager.removeView(view);
            view = null;
        }
        break;
```

说明：鉴于计算机配置的不同，且在 Android4.2 版本的模拟器上运行时，如果不能够正常显示出归属地的吐司，建议在手机接通通话时不要将吐司移除。

到此，来电归属地信息显示的功能基本实现完毕。细心的读者可能会发现一个 bug：在"设置中心"中，设置来电归属地是否开启时，我们并没有对是否开启状态进行存储，如果开启该服务，当应用程序退出时就不能够正确地显示。如果通过 sp 进行存储，一样会出现 bug（将其开启的状态存储后，用户不退出应用，然后进入应用程序管理器中将程序停止，此时显示来电归属地信息的服务也就终止了，而 sp 中此时保存的是 true）。

解决 bug 的方案：可以通过 ActivityManager 对象，来动态地获取当前正在运行的服务有哪些，如果指定的服务是运行的状态，就勾选上对应的 Checkbox；如果指定的服务没有正在运行，就取消 Checkbox 的勾选。通过 ActivityManager 可以查看服务是否正处于运行状态。我们将判断一个服务是否处于运行装载作为一个工具类，该工具类位于包"com.guoshisp.mobilesafe.utils"下。

ServiceStatusUtil.java 代码如下：

```java
package com.guoshisp.mobilesafe.utils;
import java.util.List;
import android.app.ActivityManager;
import android.app.ActivityManager.RunningServiceInfo;
import android.content.Context;
public class ServiceStatusUtil {
    /**
     * 判断某个服务是否处于运行状态
     * @param context
     * @param serviceClassName 服务的完整的类名
     * @return true 表示正在运行，false 表示没有运行
     */
    public static boolean isServiceRunning(Context context,String
                                    serviceClassName){
        ActivityManager am = (ActivityManager) context.getSystemService
                    (Context.ACTIVITY_SERVICE);
        //参数100：表示要获取正在运行的100个服务。如果没有100，则返回所有正在运行的；如果超过100，则只返回100个
        List<RunningServiceInfo> infos = am.getRunningServices(100);
        //遍历返回的服务，判断我们查看的服务是否处于运行状态
        for(RunningServiceInfo info: infos){
        if(serviceClassName.equals(info.service.getClassName())){
            return true;
            }
        }
        return false;
    }
}
```

这里介绍一下 Android 中的两个重要管理者：PackageManager 和 ActivitManager。

Android 系统为应用管理功能提供了大量的 API。根据功能的不同，这些 API 分为两大类：PackageManager 相关和 ActivitManager 相关。

PackageManager 相关（程序管理器）是静态的。本类 API 是对所有基于加载信息的数据结

构的封装，包括以下功能：安装、卸载应用；查询 permission 相关信息；查询 Application 相关信息（application，activity，receiver，service，provider 及相应属性等）；查询已安装应用；增加、删除 permission；清除用户数据、缓存、代码段等；查询非相关的 API 需要特定的权限，具体的 API 可参考 SDK 文档。

ActivityManager 相关（任务管理器）是动态的。本类 API 是对运行时管理功能和运行时数据结构的封装，包括以下功能：激活/去激活 activity；注册/取消注册动态接收 intent；发送/取消发送 intent；activity 生命周期管理（暂停、恢复、停止、销毁等）；activity task 管理（前台→后台，后台→前台，最近 task 查询，运行时 task 查询）；激活/去激活 service；激活/去激活 provider 等；task 管理相关 API 需要特定的权限，具体 API 可参考 SDK 文档。

当"设置中心"界面刚加载进来时（执行 onResum()方法时），就立即判断对应的服务是否处于运行状态，此时的 SettingCenterActivity.java 业务代码如下：

```java
package com.guoshisp.mobilesafe;
import android.app.Activity;
import android.content.Intent;
import android.content.SharedPreferences;
import android.content.SharedPreferences.Editor;
import android.graphics.Color;
import android.os.Bundle;
import android.view.View;
import android.view.View.OnClickListener;
import android.widget.CheckBox;
import android.widget.CompoundButton;
import android.widget.CompoundButton.OnCheckedChangeListener;
import android.widget.RelativeLayout;
import android.widget.TextView;
import com.guoshisp.mobilesafe.service.ShowCallLocationService;
import com.guoshisp.mobilesafe.utils.ServiceStatusUtil;
public class SettingCenterActivity extends Activity implements
    OnClickListener {
    //用于存储自动更新是否开启的 boolean 值
    private SharedPreferences sp;
    //自动更新是否开启对应的 TextView 控件的显示文字
    private TextView tv_setting_autoupdate_status;
    //显示自动更新是否开启的勾选框
    private CheckBox cb_setting_autoupdate;
    //归属地显示控件的声明
    private TextView tv_setting_show_location_status;//显示来显归属地是否开
                                                      启的状态
    private CheckBox cb_setting_show_location;//是否开启来电归属地的 Checkbox
    private RelativeLayout rl_setting_show_location;//"来电归属地是否开启"
                                                     控件的父控件
    private Intent showLocationIntent;//开启来电归属地信息显示的意图
    @Override
```

```java
protected void onCreate(Bundle savedInstanceState) {
    setContentView(R.layout.setting_center);
    super.onCreate(savedInstanceState);
    //获取Sdcard下的config.xml文件，如果该文件不存在，那么将会自动创建该文
      件，文件的操作类型为私有类型
    sp = getSharedPreferences("config", MODE_PRIVATE);
    //标记自动更新状态是否开启对应的Checkbox控件
    cb_setting_autoupdate = (CheckBox) findViewById(R.id.cb_setting_
                        autoupdate);
    //显示当前自动更新是否开启对应的TextView控件
    tv_setting_autoupdate_status = (TextView) findViewById(R.id.tv_
                        setting_autoupdate_status);
    //初始化自动更新的ui，默认状态下是开启的
    boolean autoupdate = sp.getBoolean("autoupdate", true);
    if (autoupdate) {
        tv_setting_autoupdate_status.setText("自动更新已经开启");
        //因为autoupdate变量为true,则表示自动更新开启，所以，Checkbox的
          状态应该是勾选状态的，即为true
        cb_setting_autoupdate.setChecked(true);
    } else {
        tv_setting_autoupdate_status.setText("自动更新已经关闭");
        //因为autoupdate变量为false,则表示自动更新未开启，所以，Checkbox
          的状态应该是未勾选状态的，即为false
        cb_setting_autoupdate.setChecked(false);
    }
    /**
     * 当Checkbox的状态发生改变时执行以下代码
     */
    cb_setting_autoupdate
            .setOnCheckedChangeListener(new OnCheckedChangeListener() {
        //参数一：当前的Checkbox；第二个参数：当前的Checkbox是否处于勾选状态
            public void onCheckedChanged(CompoundButton buttonView,
                    boolean isChecked) {
                //获取编辑器
                Editor editor = sp.edit();
    //持久化存储当前Checkbox的状态，当下次进入时，依然可以保存当前设置的状态
                editor.putBoolean("autoupdate", isChecked);
                //将数据真正提交到sp里面
                editor.commit();
                if (isChecked) {//Checkbox处于选中效果
    //当Checkbox处于勾选状态时，表示自动更新已经开启，同时修改字体颜色
                    tv_setting_autoupdate_status
                            .setTextColor(Color.WHITE);
                    tv_setting_autoupdate_status.setText("自动更
                        新已经开启");
                } else {//Checkbox处于未勾选状态
```

```java
                    //当Checkbox未处于勾选状态时，表示自动更新已经开启，同时修改字体颜色
                        tv_setting_autoupdate_status
                            .setTextColor(Color.RED);
                        tv_setting_autoupdate_status.setText("自动更
                            新已经关闭");
                    }
                }
            });
        //显示归属地信息的ui初始化
        tv_setting_show_location_status = (TextView) findViewById(R.id.tv_
                            setting_show_location_status);
        cb_setting_show_location = (CheckBox) findViewById(R.id.cb_setting_
                            show_location);
        rl_setting_show_location = (RelativeLayout) findViewById(R.id.rl_
                            setting_show_location);
        showLocationIntent = new Intent(this, ShowCallLocationService.class);
        rl_setting_show_location.setOnClickListener(this);
    }
@Override
protected void onResume() {
    if(ServiceStatusUtil.isServiceRunning(this,"com.guoshisp.mobilesafe.
       service.ShowCallLocationService")) {
            cb_setting_show_location.setChecked(true);
            tv_setting_show_location_status.setText("来电归属地显示已经开启");
    } else {
            cb_setting_show_location.setChecked(false);
            tv_setting_show_location_status.setText("来电归属地显示没有开启");
    }
        super.onResume();
}
    //响应单击事件
    @Override
    public void onClick(View v) {
        //TODO Auto-generated method stub
        switch (v.getId()) {
        case R.id.rl_setting_show_location://来电归属地是否开启
            if (cb_setting_show_location.isChecked()) {
                tv_setting_show_location_status.setText("来电归属地显示没
                                                        有开启");
                stopService(showLocationIntent);
                cb_setting_show_location.setChecked(false);
            } else {
                tv_setting_show_location_status.setText("来电归属地显示已
                                                        经开启");
                startService(showLocationIntent);
                cb_setting_show_location.setChecked(true);
```

                }
                break;
            }
        }
    }

以上基本上实现了来电归属地信息显示的功能，我们也可以模仿该功能来实现外拨电话的归属地信息显示的功能。

外拨电话显示归属地信息的实现思路：在讲解"手机防盗"模块时，我们通过广播来拦截一个外拨电话来实现快速进入"手机防盗"界面。所以，我们可以考虑将来电归属地中的核心代码粘贴过来。但是有一点是需要考虑的：广播的生命周期比较短（不会超过 20 秒），当超过 20 秒后，windowManager.addView(view, params)中的 view 就会被回收，此时程序就会报错。所以，我们可以在该广播中开启一个服务来实现。有兴趣的读者可以尝试一下。

3.4 更改归属地的显示风格

需求：在设置中心单击"来电归属地风格设置"条目后，弹出一个对话框让我们选择归属地提示框的风格，当选中一种风格后，来电归属地信息显示的吐司背景色就会发生对应的改变。

实现思路：在设置中心中设置显示风格后，条目的底部也会显示出风格的名称，同时再将该风格存入 Item 对应的 id，存储在 sp 中，在显示来电归属地的服务中取出 sp 中对应的 id 后，来为其设置对应风格的背景图片。

设置中心对应的 SettingCenterActivity.java 的业务代码如下：

```
package com.guoshisp.mobilesafe;
import android.app.Activity;
import android.app.AlertDialog;
import android.app.AlertDialog.Builder;
import android.content.DialogInterface;
import android.content.Intent;
import android.content.SharedPreferences;
import android.content.SharedPreferences.Editor;
import android.graphics.Color;
import android.os.Bundle;
import android.view.View;
import android.view.View.OnClickListener;
import android.widget.CheckBox;
import android.widget.CompoundButton;
import android.widget.CompoundButton.OnCheckedChangeListener;
import android.widget.RelativeLayout;
import android.widget.TextView;
import com.guoshisp.mobilesafe.service.ShowCallLocationService;
```

```java
import com.guoshisp.mobilesafe.utils.ServiceStatusUtil;
public class SettingCenterActivity extends Activity implements
    OnClickListener {
    //程序的自动更新
    private SharedPreferences sp;//用于存储自动更新是否开启的boolean值
    private TextView tv_setting_autoupdate_status;//自动更新是否开启对应的
                                                  TextView控件的显示文字
    private CheckBox cb_setting_autoupdate;//显示自动更新是否开启的勾选框
    //归属地显示控件的声明
    private TextView tv_setting_show_location_status;//显示来显归属地是否开
                                                      启的状态
    private CheckBox cb_setting_show_location;//是否开启来电归属地的Checkbox
    private RelativeLayout rl_setting_show_location;//"来电归属地是否开启"
                                                     控件的父控件
    private Intent showLocationIntent;//开启来电归属地信息显示的意图
    //归属地显示背景控件的声明
    private RelativeLayout rl_setting_change_bg;//"来电归属地风格设置"控件
                                                 的父控件
    private TextView tv_setting_show_bg;//"来电归属地风格设置"下用于显示当前
                                         的风格文字
    @Override
    protected void onCreate(Bundle savedInstanceState) {
        setContentView(R.layout.setting_center);
        super.onCreate(savedInstanceState);
        //获取Sdcard下的config.xml文件,如果该文件不存在,那么将会自动创建该文
          件,文件的操作类型为私有类型
        sp = getSharedPreferences("config", MODE_PRIVATE);
        //标记自动更新状态是否开启对应的Checkbox控件
        cb_setting_autoupdate = (CheckBox) findViewById(R.id.cb_setting_
                            autoupdate);
        //显示当前自动更新是否开启对应的TextView控件
        tv_setting_autoupdate_status = (TextView) findViewById(R.id.tv_
                            setting_autoupdate_status);
        //初始化自动更新的ui,默认状态下是开启的
        boolean autoupdate = sp.getBoolean("autoupdate", true);
        if (autoupdate) {
            tv_setting_autoupdate_status.setText("自动更新已经开启");
            //因为autoupdate变量为true,则表示自动更新开启,所以,Checkbox的
              状态应该是勾选状态的,即为true
            cb_setting_autoupdate.setChecked(true);
        } else {
            tv_setting_autoupdate_status.setText("自动更新已经关闭");
            //因为autoupdate变量为false,则表示自动更新未开启,所以,Checkbox
              的状态应该是未勾选状态的,即为false
            cb_setting_autoupdate.setChecked(false);
        }
```

```java
        /**
         * 当 Checkbox 的状态发生改变时执行以下代码
         */
        cb_setting_autoupdate
                .setOnCheckedChangeListener(new OnCheckedChangeListener() {
                    //参数一：当前的 Checkbox；第二个参数：当前的 Checkbox 是否处于勾选状态
                    public void onCheckedChanged(CompoundButton buttonView,
                            boolean isChecked) {
                        //获取编辑器
                        Editor editor = sp.edit();
                        //持久化存储当前 Checkbox 的状态，当下次进入时，依然可以保存当前设置的状态
                        editor.putBoolean("autoupdate", isChecked);
                        //将数据真正提交到 sp 里面
                        editor.commit();
                        if (isChecked) {//Checkbox 处于选中效果
                            //当 Checkbox 处于勾选状态时，表示自动更新已经开启，同时修改字体颜色
                            tv_setting_autoupdate_status
                                    .setTextColor(Color.WHITE);
                            tv_setting_autoupdate_status.setText("自动更新已经开启");
                        } else {//Checkbox 处于未勾选状态
                            //当 Checkbox 未处于勾选状态时，表示自动更新已经开启，同时修改字体颜色
                            tv_setting_autoupdate_status
                                    .setTextColor(Color.RED);
                            tv_setting_autoupdate_status.setText("自动更新已经关闭");
                        }
                    }
                });
        //显示归属地信息的 ui 初始化
        tv_setting_show_location_status = (TextView) findViewById(R.id.tv_
                                setting_show_location_status);
        cb_setting_show_location = (CheckBox) findViewById(R.id.cb_setting_
                                show_location);
        rl_setting_show_location = (RelativeLayout) findViewById(R.id.rl_
                                setting_show_location);
        showLocationIntent = new Intent(this, ShowCallLocationService.class);
        rl_setting_show_location.setOnClickListener(this);
        //归属地显示背景的声明
        rl_setting_change_bg = (RelativeLayout) findViewById(R.id.rl_
                                setting_change_bg);
        tv_setting_show_bg = (TextView) findViewById(R.id.tv_setting_
                                show_bg);
        rl_setting_change_bg.setOnClickListener(this);
    }
    @Override
```

```java
protected void onResume() {
    if(ServiceStatusUtil.isServiceRunning(this,"com.guoshisp.mobilesafe.
       service.ShowCallLocationService")) {
        cb_setting_show_location.setChecked(true);
        tv_setting_show_location_status.setText("来电归属地显示已经开启");
    } else {
        cb_setting_show_location.setChecked(false);
        tv_setting_show_location_status.setText("来电归属地显示没有开启");
    }
    super.onResume();
}
//响应单击事件
@Override
public void onClick(View v) {
    //TODO Auto-generated method stub
    switch (v.getId()) {
    case R.id.rl_setting_show_location://来电归属地是否开启
        if (cb_setting_show_location.isChecked()) {
            tv_setting_show_location_status.setText("来电归属地显示没有开启");
            stopService(showLocationIntent);
            cb_setting_show_location.setChecked(false);
        } else {
            tv_setting_show_location_status.setText("来电归属地显示已经开启");
            startService(showLocationIntent);
            cb_setting_show_location.setChecked(true);
        }
        break;
    case R.id.rl_setting_change_bg://来电归属地风格设置
        showChooseBgDialog();
        break;
    }
}
/**
 * 更改背景颜色的对话框
 */
private void showChooseBgDialog() {
    //获取一个对话框构造器
    AlertDialog.Builder builder = new Builder(this);
    //设置对话框标题的图标
    builder.setIcon(R.drawable.notification);
    //设置对话框的标题
    builder.setTitle("归属地提示框风格");
    //对话框中item的对应显示文字
    final String[] items = { "半透明", "活力橙", "卫士蓝", "苹果绿", "金
                              属灰" };
    //用于显示对话框中那一个条目被选中。默认的是第一个条目
    int which = sp.getInt("which", 0);
```

```java
            //设置单个选择条目。Item 中，只能有一个处于选中状态
            builder.setSingleChoiceItems(items, which,
                    new DialogInterface.OnClickListener() {
                        //处理 Item 的单击事件
                        public void onClick(DialogInterface dialog, int which) {
                            //将条目的 id 存入 sp 中
                            Editor editor = sp.edit();
                            editor.putInt("which", which);
                            editor.commit();
                            //设置 Item 的文字信息
                            tv_setting_show_bg.setText(items[which]);
                            //关闭对话框
                            dialog.dismiss();
                        }
                    });
            builder.setNegativeButton("取消", new DialogInterface.
                            OnClickListener() {
                public void onClick(DialogInterface dialog, int which) {
                }
            });
            //创建并显示出对话框
            builder.create().show();
        }
    }
```

说明：setContentView(R.layout.setting_center)中的布局文件在 3.3 节中已经给出。

当电话到来时，应当在"来电归属地显示"的服务中完成对吐司的背景色的设置，ShowCallLocationService.java 对应的业务代码如下：

```java
package com.guoshisp.mobilesafe.service;
import android.app.Service;
import android.content.Intent;
import android.content.SharedPreferences;
import android.graphics.PixelFormat;
import android.os.IBinder;
import android.telephony.PhoneStateListener;
import android.telephony.TelephonyManager;
import android.view.View;
import android.view.WindowManager;
import android.view.WindowManager.LayoutParams;
import android.widget.LinearLayout;
import android.widget.TextView;
import com.guoshisp.mobilesafe.R;
import com.guoshisp.mobilesafe.db.dao.NumberAddressDao;
//在后台监听电话呼入的状态
public class ShowCallLocationService extends Service {
    private TelephonyManager tm;//电话管理器
    private MyPhoneListener listener;//电话状态改变的监听器
```

```java
private WindowManager windowManager;//窗体管理器
private SharedPreferences sp;//用于取出归属地风格显示风格的Item对应的id
//"半透明","活力橙","卫士蓝","苹果绿","金属灰"
private static final int[] bgs = {R.drawable.call_locate_white,R.
                                  drawable.call_locate_orange,
R.drawable.call_locate_blue,R.drawable.call_locate_green,R.drawable.
  call_locate_gray};
@Override
public IBinder onBind(Intent intent) {
    return null;
}
/**
 * 当服务第一次被创建时调用
 */
@Override
public void onCreate() {
    super.onCreate();
    sp =getSharedPreferences("config",MODE_PRIVATE);
    //注册系统的电话状态改变的监听器
    listener = new MyPhoneListener();
    //获取系统的电话管理器
    tm = (TelephonyManager) getSystemService(TELEPHONY_SERVICE);
    //为电话设置一个监听。参数一：监听器；参数二：要监听的电话改变类型（这里监听
      的是通话状态）
    tm.listen(listener, PhoneStateListener.LISTEN_CALL_STATE);
    windowManager = (WindowManager) getSystemService(WINDOW_SERVICE);
}
private class MyPhoneListener extends PhoneStateListener {
    private View view;
    //参数一：手机的状态；参数二：呼叫进来的手机号码
    @Override
    public void onCallStateChanged(int state, String incomingNumber) {
        switch (state) {
        case TelephonyManager.CALL_STATE_RINGING://手机铃声正在响
            //获取呼叫进来号码的地址（查询之前的号码归属地数据库）
            String address = NumberAddressDao.getAddress
                        (incomingNumber);
            //使用系统的吐司来显示归属地信息，但显示的时间较短
            //Toast.makeText(getApplicationContext(), "归属地:"+address,
              1).show();
            //通过布局填充器将一个显示号码归属地的布局转成View,该View是一个吐司
            view = View.inflate(getApplicationContext(), R.layout.show_
                        address, null);
            //获取到显示号码归属地布局的根布局LinearLayout
            LinearLayout ll = (LinearLayout) view.findViewById(R.id.
                        ll_show_address);
            //从sp文件中获取显示归属地风格的Item的id
```

```
            int which = sp.getInt("which", 0);
            //设置来电归属地显示的背景图片
            ll.setBackgroundResource(bgs[which]);
            //查找 view 中的用于显示归属地的 TextView
            TextView tv = (TextView) view.findViewById(R.id.tv_show_
                        address);
            //将归属地信息设置到 TextView
            tv.setText(address);
//获取到与窗体相关的布局的参数（这里用于设置窗体上显示来电归属地的吐司的参数信息）
            final WindowManager.LayoutParams params = new LayoutParams();
            //设置窗体布局 View 的高度
            params.height = WindowManager.LayoutParams.WRAP_CONTENT;
            //设置窗体布局 View 的宽度
            params.width = WindowManager.LayoutParams.WRAP_CONTENT;
            //窗体 View 不可以获取焦点，不可以被触摸、保持在屏幕上
            params.flags = WindowManager.LayoutParams.FLAG_NOT_FOCUSABLE
                | WindowManager.LayoutParams.FLAG_NOT_TOUCHABLE
                | WindowManager.LayoutParams.FLAG_KEEP_SCREEN_ON;
            //显示在窗体上的 style 为半透明
            params.format = PixelFormat.TRANSLUCENT;
            //窗体 View 的类型为吐司
            params.type = WindowManager.LayoutParams.TYPE_TOAST;
            //将吐司挂载在窗体上。窗体服务是一个全局的系统服务，该服务开启后会在后
              台运行。一般情况下，在窗体上一旦挂载一个 View 并显示后，并不会自动消失
            windowManager.addView(view, params);
            break;
        case TelephonyManager.CALL_STATE_IDLE: //手机的空闲状态
            if(view!=null){
                //将窗体上的吐司移除
                windowManager.removeView(view);
                view = null;
            }
            break;
        case TelephonyManager.CALL_STATE_OFFHOOK://手机接通通话时的状态
            if(view!=null){
                //将窗体上的吐司移除
                windowManager.removeView(view);
                view = null;
            }
            break;
        }
        super.onCallStateChanged(state, incomingNumber);
    }
}
/**
 * 取消电话状态的监听
 */
```

```
@Override
public void onDestroy() {
    super.onDestroy();
    tm.listen(listener, PhoneStateListener.LISTEN_NONE);
    listener = null;
}
}
```

测试运行：默认情况下使用之前设置的半透明的颜色。进入"设置中心"，开启来电归属地设置，单击"来电归属地风格设置"条目后会弹出一个窗体让我们选择其中的风格，选择后，对话框消失，此时的设置中心的界面如图3-11所示。呼入一个手机号码后，归属地的显示风格如图3-12所示。

图 3-11

图 3-12

3.5 更改归属地的显示位置

需求：在进入设置中心单击"归属地提示框的位置"时，进入一个设置归属地位置显示的界面，该界面由两部分组成——一个是类似于吐司的View（View上显示的文字是：双击居中），该View用于显示归属地信息，可以在屏幕上移动该View，并且双击该View后，View会在当前位置的水平方向上居中显示，View 最终被拖动到屏幕的某个地方后，来电归属地的显示位置也会跟着改变；另一个是一个提示框（提示内容是：按住提示框拖动任意位置，按手机返回键立刻生效）。另外，当View被移动到手机屏幕的下半部分时，提示框会跳到上半部分显示；当View被移动到手机屏幕的上半部分时，提示框会跳到下半部分显示。

实现思路：解决这样一个需求的核心点在于控制View的移动。实现View在屏幕上的移动

需要以下几步:

(1) 记录下手指第一次触动到 View 的坐标。

(2) 计算出开始的触摸点到结束时的触摸点在 X 轴和 Y 轴的移动距离。

(3) 立即更改 View 在窗体中的位置:水平方向平移 X 个距离,垂直平移 Y 个距离。

(4) 立即将手指第一次触摸屏幕的位置更新为当前的位置,以便下次继续移动。

首先,进入设置中心为"归属地提示框的位置"条目设置单击事件,SettingCenterActivity.java 的业务代码见在线资源包中代码文本部分 3.5.doc。

当单击"归属地提示框的位置"时,将开启一个新的 Activity,该 Activity 用于实现归属地信息显示位置的更改,DragViewActivity.java 的业务代码如下:

```java
package com.guoshisp.mobilesafeafe;
import android.app.Activity;
import android.os.Bundle;
import android.util.Log;
import android.view.MotionEvent;
import android.view.View;
import android.view.View.OnTouchListener;
import android.widget.ImageView;
import android.widget.TextView;
import com.guoshisp.mobilesafe.R;
public class DragViewActivity extends Activity {
    protected static final String TAG = "DragViewActivity";
    private ImageView iv_drag_view;//要移动的 View
    private TextView tv_drag_view;//提示框
    @Override
    protected void onCreate(Bundle savedInstanceState) {
        super.onCreate(savedInstanceState);
        setContentView(R.layout.drag_view);
        iv_drag_view = (ImageView) findViewById(R.id.iv_drag_view);
        tv_drag_view = (TextView) findViewById(R.id.tv_drag_view);
        //为 View 注册一个被触摸事件的监听器
        iv_drag_view.setOnTouchListener(new OnTouchListener() {
            //记录起始触摸点的坐标
            int startx;//记录起始时的 X 坐标
            int starty;//记录起始时的 Y 坐标
            public boolean onTouch(View v, MotionEvent event) {
                switch (event.getAction()) {
                case MotionEvent.ACTION_DOWN://手指第一次接触屏幕
                    Log.i(TAG, "摸到");
                    startx = (int) event.getRawX();//获取到手指触摸点的 X 坐标
                    starty = (int) event.getRawY();//获取到手指触摸点的 Y 坐标
                    break;
                case MotionEvent.ACTION_MOVE://手指在屏幕上移动
```

```java
                    int x = (int) event.getRawX();//获取到当前手指触摸点的X坐标
                    int y = (int) event.getRawY();//获取到当前手指触摸点的Y坐标
                    int dx = x - startx;//计算出View在屏幕X轴方向上被移动
                                        的距离
                    int dy = y - starty;//计算出View在屏幕Y轴方向上被移动
                                        的距离
                    //计算出被拖动的View距离窗体上、下、左、右的距离
                    int t = iv_drag_view.getTop();
                    int b = iv_drag_view.getBottom();
                    int l = iv_drag_view.getLeft();
                    int r = iv_drag_view.getRight();
                    //获取到移动后的View的在窗体中的位置
                    int newl = l+dx;
                    int newt = t+dy;
                    int newr = r+dx;
                    int newb = b+dy;
                    //将移动后的View在窗体上重新显示出来
                    iv_drag_view.layout(newl,newt,newr, newb);
                    //立即更新手指第一次触摸屏幕的位置坐标,以便下次继续移动
                    startx = (int) event.getRawX();
                    starty = (int) event.getRawY();
                    Log.i(TAG, "移动");
                    break;
                case MotionEvent.ACTION_UP://手指离开屏幕
                    Log.i(TAG, "松手");
                    break;
                }
                //true 会消费调当前的触摸事件,那么后面的移动和离开事件会被响应到
                //false 不会消费当前的触摸事件,那么后面的移动和离开事件都不会被响应到
                return true;
            }
        });
    }
}
```

测试运行:首先在清单文件中为 DragViewActivity 做好配置信息。运行程序后,进入到设置中心单击"归属地提示框的位置"所对应的条目,界面如图 3-13 所示。在模拟器上,用鼠标左键单击并长按鼠标左键便可拖动 View,将 View 拖动到屏幕的右下方,界面效果图如图 3-14 所示。此时的 DragViewActivity.java 中的业务代码如下:

```java
package com.guoshisp.mobilesafe;
import android.app.Activity;
import android.os.Bundle;
import android.util.Log;
import android.view.MotionEvent;
import android.view.View;
import android.view.View.OnTouchListener;
import android.widget.ImageView;
```

```java
import android.widget.TextView;
import com.guoshisp.mobilesafe.R;
public class DragViewActivity extends Activity {
    protected static final String TAG = "DragViewActivity";
    private ImageView iv_drag_view;//要移动的View
    private TextView tv_drag_view;//提示框
    @Override
    protected void onCreate(Bundle savedInstanceState) {
        super.onCreate(savedInstanceState);
        setContentView(R.layout.drag_view);
        iv_drag_view = (ImageView) findViewById(R.id.iv_drag_view);
        tv_drag_view = (TextView) findViewById(R.id.tv_drag_view);
        //为View注册一个被触摸事件的监听器
        iv_drag_view.setOnTouchListener(new OnTouchListener() {
            //记录起始触摸点的坐标
            int startx;//记录起始时的X坐标
            int starty;//记录起始时的Y坐标
            public boolean onTouch(View v, MotionEvent event) {
                switch (event.getAction()) {
                case MotionEvent.ACTION_DOWN://手指第一次接触屏幕
                    Log.i(TAG, "摸到");
                    startx = (int) event.getRawX();//获取到手指触摸点的X坐标
                    starty = (int) event.getRawY();//获取到手指触摸点的Y坐标
                    break;
                case MotionEvent.ACTION_MOVE://手指在屏幕上移动
                    int x = (int) event.getRawX();//获取到当前手指触摸点的X坐标
                    int y = (int) event.getRawY();//获取到当前手指触摸点的Y坐标
                    //获取提示框的高度
                    int tv_height = tv_drag_view.getBottom() - tv_drag_
                            view.getTop();
                    //判断View是处于窗体的上方还是下方
                    if(y>(windowHeight/2)){//手指移动到了窗体的下半部分
                        //将提示框移动到窗体的上半部分
                        tv_drag_view.layout(tv_drag_view.getLeft(), 60,
                            tv_drag_view.getRight(), 60+tv_height);
                    }else{//手指移动到了窗体的上半部分
                        //将提示框移动到窗体的下半部分
                        tv_drag_view.layout(tv_drag_view.getLeft(),
windowHeight-20-tv_height, tv_drag_view.getRight(), windowHeight-20);
                    }
                    int dx = x - startx;//计算出View在屏幕X轴方向上被移动
                                        的距离
                    int dy = y - starty;//计算出View在屏幕Y轴方向上被移动
                                        的距离
                    //计算出被拖动的View距离窗体上、下、左、右的距离
                    int t =iv_drag_view.getTop();
                    int b = iv_drag_view.getBottom();
                    int l = iv_drag_view.getLeft();
```

```
                int r = iv_drag_view.getRight();
                //获取到移动后的View的在窗体中的位置
                int newl = l+dx;
                int newt = t+dy;
                int newr = r+dx;
                int newb = b+dy;
                //将移动后的View在窗体上重新显示出来
                iv_drag_view.layout(newl,newt,newr, newb);
                //立即更新手指第一次触摸屏幕的位置坐标，以便下次继续移动
                startx = (int) event.getRawX();
                starty = (int) event.getRawY();
                Log.i(TAG, "移动");
                break;
            case MotionEvent.ACTION_UP://手指离开屏幕
                Log.i(TAG, "松手");
                break;
            }
            //true 会消费调当前的触摸事件，那么后面的移动和离开事件会被响应到
            //false 不会消费当前的触摸事件，那么后面的移动和离开事件都不会被响应到
            return true;
        }
    });
}
```

图 3-13

图 3-14

说明：触摸事件的方法 onTouch(View v, MotionEvent event)的返回值是一个 boolean 值。当返回值为 true 时，会消费当前的触摸事件，那么后面的移动和离开事件会被响应到；当返回值为 false 时，不会消费当前的触摸事件，那么后面的移动和离开事件都不会被响应到（View 也就不会被移动）。可以通过打印 Log 来测试。

上面实现了 View 的拖动,下面需要实现的是:当 View 移动到窗体的下半部分时,提示框在窗体的上半部分显示;当 View 移动到窗体的上半部分时,提示框在窗体的下半部分显示。并且使用 sp 记录住当前 View 的显示位置,当下次进入时显示的是上次触摸结束后的位置。

实现思路:首先需要获取窗体的管理者来获取当前窗体的高度,然后通过计算当前 View 所在的位置与屏幕的二分之一高度进行比较后,将提示框移动到对应的位置。

此时,DragViewActivity.java 的业务代码如下:

```java
package com.guoshisp.mobilesafe;
import android.app.Activity;
import android.content.SharedPreferences;
import android.content.SharedPreferences.Editor;
import android.os.Bundle;
import android.util.Log;
import android.view.MotionEvent;
import android.view.View;
import android.view.View.OnTouchListener;
import android.widget.ImageView;
import android.widget.RelativeLayout;
import android.widget.RelativeLayout.LayoutParams;
import android.widget.TextView;
public class DragViewActivity extends Activity {
    protected static final String TAG = "DragViewActivity";
    private ImageView iv_drag_view;//要移动的View
    private TextView tv_drag_view;//提示框
    private int windowHeight;//定义屏幕的高度
    private int windowWidth;//定义屏幕的宽度
    private SharedPreferences sp;//用于存储View的位置信息
    @Override
    protected void onCreate(Bundle savedInstanceState) {
        super.onCreate(savedInstanceState);
        setContentView(R.layout.drag_view);
        iv_drag_view = (ImageView) findViewById(R.id.iv_drag_view);
        tv_drag_view = (TextView) findViewById(R.id.tv_drag_view);
        windowHeight = getWindowManager().getDefaultDisplay().getHeight();
        windowWidth = getWindowManager().getDefaultDisplay().getWidth();
        sp = getSharedPreferences("config", MODE_PRIVATE);
        //初始化上次移动后的View的显示位置
        RelativeLayout.LayoutParams params = (LayoutParams) iv_drag_view.
                                                getLayoutParams();
        params.leftMargin = sp.getInt("lastx", 0);//获取到被移动的View离窗
                                                    体左端的X值
        params.topMargin = sp.getInt("lasty", 0);//获取到被移动的View离窗体
                                                    顶端的Y值
        iv_drag_view.setLayoutParams(params);
```

```java
//为View注册一个被触摸事件的监听器
iv_drag_view.setOnTouchListener(new OnTouchListener() {
    //记录起始触摸点的坐标
    int startx;//记录起始时的X坐标
    int starty;//记录起始时的Y坐标
    public boolean onTouch(View v, MotionEvent event) {
        switch (event.getAction()) {
            case MotionEvent.ACTION_DOWN://手指第一次接触屏幕
                Log.i(TAG, "摸到");
                startx = (int) event.getRawX();//获取到手指触摸点的X坐标
                starty = (int) event.getRawY();//获取到手指触摸点的Y坐标
                break;
            case MotionEvent.ACTION_MOVE://手指在屏幕上移动
                int x = (int) event.getRawX();//获取到当前手指触摸点的X坐标
                int y = (int) event.getRawY();//获取到当前手指触摸点的Y坐标
                //获取提示框的高度
                int tv_height = tv_drag_view.getBottom() - tv_drag_view.getTop();
                //判断View是处于窗体的上方还是下方
                if(y>(windowHeight/2)){//手指移动到了窗体的下半部分
                    //将提示框移动到窗体的上半部分。四个参数分别为提示框距离窗体的左、上、右、下端的距离
                    tv_drag_view.layout(tv_drag_view.getLeft(), 60, tv_drag_view.getRight(), 60+tv_height);
                }else{//手指移动到了窗体的上半部分
                    //将提示框移动到窗体的下半部分
                    tv_drag_view.layout(tv_drag_view.getLeft(), windowHeight-20-tv_height, tv_drag_view.getRight(), windowHeight-20);
                }
                int dx = x - startx;//计算出View在屏幕X轴方向上被移动的距离
                int dy = y - starty;//计算出View在屏幕Y轴方向上被移动的距离
                //计算出被拖动的View距离窗体上、下、左、右的距离
                int t =iv_drag_view.getTop();
                int b = iv_drag_view.getBottom();
                int l = iv_drag_view.getLeft();
                int r = iv_drag_view.getRight();
                //获取到移动后的View的在窗体中的位置
                int newl = l+dx;
                int newt = t+dy;
                int newr = r+dx;
                int newb = b+dy;
                //将移动后的View在窗体上重新显示出来
                iv_drag_view.layout(newl,newt,newr, newb);
                //立即更新手指第一次触摸屏幕的位置坐标,以便下次继续移动
```

```
                    startx = (int) event.getRawX();
                    starty = (int) event.getRawY();
                    Log.i(TAG, "移动");
                    break;
                case MotionEvent.ACTION_UP://手指离开屏幕
                    Log.i(TAG, "松手");
//记录当前imageview在窗体中的位置（左上角的顶点距离屏幕的宽度和高度）
                    Editor editor = sp.edit();
                    int lasty = iv_drag_view.getTop();
                    int lastx = iv_drag_view.getLeft();
                    editor.putInt("lastx", lastx);
                    editor.putInt("lasty", lasty);
                    editor.commit();
                    break;
                }
//true 会消费调当前的触摸事件，那么后面的移动和离开事件会被响应到
//false 不会消费当前的触摸事件，那么后面的移动和离开事件都不会被响应到
                return true;
            }
        });
    }
}
```

测试运行：当 View 位于屏幕的上半部分时，提示框应当显示在屏幕的下半部分。界面效果图如图 3-15 所示。

但是该界面中的 View 在被移动时可以移动到窗体以外，如果将 View 的一半移出窗体，然后再次进入该界面时，View 的大小会变为原来的一半（文字和图片正常显示），如图 3-16 所示。

图 3-15　　　　　　　　　　　　　图 3-16

产生该问题的原因在于：当 View 被移出一半后，此时 View 的位置被存入到 sp 中，当下次从 sp 中取出数据并显示 View 中的图片和文字时，窗体会发现 View 的空间不够，窗体对象会自动将 View 中的图片和文字以对应的比例显示在剩余的 View 上。为了避免出现这样的问题，我们可以通过移动后的 View 距离窗体的位置来控制其移动、显示的范围。下面代码中，在解决该问题的同时，还实现了 View 双击后水平居中的效果。

实现 View 水平居中显示的思路：首先计算出 View 的宽度一半的宽度为 X，然后计算出窗体宽度的一半的宽度为 Y。

> View 居中时离窗体左边边框的距离 = Y-X。

> View 居中时离窗体右边边框的距离 = Y+X。

```java
package com.guoshisp.mobilesafe;
import android.app.Activity;
import android.content.SharedPreferences;
import android.content.SharedPreferences.Editor;
import android.os.Bundle;
import android.util.Log;
import android.view.MotionEvent;
import android.view.View;
import android.view.View.OnClickListener;
import android.view.View.OnTouchListener;
import android.widget.ImageView;
import android.widget.RelativeLayout;
import android.widget.RelativeLayout.LayoutParams;
import android.widget.TextView;
public class DragViewActivity extends Activity {
    protected static final String TAG = "DragViewActivity";
    private ImageView iv_drag_view;//要移动的View
    private TextView tv_drag_view;//提示框
    private int windowHeight;//定义屏幕的高度
    private int windowWidth;//定义屏幕的宽度
    private SharedPreferences sp;//用于存储View的位置信息
    private long firstclicktime;//记录"双击居中"时的第一次单击时间,记录的原因
                                 在于判断是否属于双击事件

    @Override
    protected void onCreate(Bundle savedInstanceState) {
        super.onCreate(savedInstanceState);
        setContentView(R.layout.drag_view);
        iv_drag_view = (ImageView) findViewById(R.id.iv_drag_view);
        tv_drag_view = (TextView) findViewById(R.id.tv_drag_view);
        windowHeight = getWindowManager().getDefaultDisplay().getHeight();
        windowWidth = getWindowManager().getDefaultDisplay().getWidth();
        sp = getSharedPreferences("config", MODE_PRIVATE);
        //初始化上次移动后的View的显示位置
```

```
RelativeLayout.LayoutParams params = (LayoutParams) iv_drag_view
        .getLayoutParams();
params.leftMargin = sp.getInt("lastx", 0);//获取到被移动的View离窗
                                            体左端的X值
params.topMargin = sp.getInt("lasty", 0);//获取到被移动的View离窗体
                                            顶端的Y值
iv_drag_view.setLayoutParams(params);
//处理View双击居中的单击事件
iv_drag_view.setOnClickListener(new OnClickListener() {
    public void onClick(View v) {
        Log.i(TAG, "我被单击啦......................");
        //判断是第一次单击，还是第二次单击
        if (firstclicktime > 0) {//第二次的单击事件。因为
            firstclicktime是一个成员变量，默认值为0
            long secondclickTime = System.currentTimeMillis();
            if (secondclickTime - firstclicktime < 500) {//设定双
                击的阈值为0.5秒
                Log.i(TAG, "双击啦......................");
                //双击后，需要将第一次的单击时间设置为0，以便下次单击
                firstclicktime = 0;
                //计算出View的宽度
                int right = iv_drag_view.getRight();
                int left = iv_drag_view.getLeft();
                int iv_width = right - left;//计算出View的长度
                //计算出View在窗体正中央时的View左端离窗体左边边框的
                    距离和View右端离窗体右边边框的距离
                int iv_left = windowWidth / 2 - iv_width / 2;
                int iv_right = windowWidth / 2 + iv_width / 2;
                //将View显示到界面的最中央.
                iv_drag_view.layout(iv_left, iv_drag_view.getTop(),
                        iv_right, iv_drag_view.getBottom());
                //将View在中央显示的位置数据存入sp中
                Editor editor = sp.edit();
                int lasty = iv_drag_view.getTop();
                int lastx = iv_drag_view.getLeft();
                editor.putInt("lastx", lastx);
                editor.putInt("lasty", lasty);
                editor.commit();
            }
        }
        firstclicktime = System.currentTimeMillis();
        //解决用户的奇怪操作：单击一下停留较长，然后双击
        new Thread() {
            public void run() {
```

```java
                    try {
                        Thread.sleep(500);
                        firstclicktime = 0;
                    } catch (InterruptedException e) {
                        e.printStackTrace();
                    }
                }
            };
        }.start();
    }
});
//为View注册一个被触摸事件的监听器
iv_drag_view.setOnTouchListener(new OnTouchListener() {
    //记录起始触摸点的坐标
    int startx;//记录起始时的X坐标
    int starty;//记录起始时的Y坐标
    public boolean onTouch(View v, MotionEvent event) {
        switch (event.getAction()) {
        case MotionEvent.ACTION_DOWN://手指第一次接触屏幕
            Log.i(TAG, "摸到");
            startx = (int) event.getRawX();//获取到手指触摸点的X坐标
            starty = (int) event.getRawY();//获取到手指触摸点的Y坐标
            break;
        case MotionEvent.ACTION_MOVE://手指在屏幕上移动
            int x = (int) event.getRawX();//获取到当前手指触摸点的X坐标
            int y = (int) event.getRawY();//获取到当前手指触摸点的Y坐标
            //获取提示框的高度
            int tv_height = tv_drag_view.getBottom()
                    - tv_drag_view.getTop();
            //判断View是处于窗体的上方还是下方
            if (y > (windowHeight / 2)) {//手指移动到了窗体的下一半
                //将提示框移动到窗体的上半部分。四个参数分别为提示框距离窗体的左、上、右、下端的距离
                tv_drag_view.layout(tv_drag_view.getLeft(), 60,
                        tv_drag_view.getRight(), 60 + tv_height);
            } else {//手指移动到了窗体的上半部分
                //将提示框移动到窗体的下半部分
                tv_drag_view.layout(tv_drag_view.getLeft(),
                        windowHeight - 20 - tv_height,
                        tv_drag_view.getRight(), windowHeight - 20);
            }
            int dx = x - startx;//计算出View在屏幕X轴方向上被移动
                                //的距离
            int dy = y - starty;//计算出View在屏幕Y轴方向上被移动
                                //的距离
            //计算出被拖动的View距离窗体上、下、左、右的距离
```

```java
                    int t = iv_drag_view.getTop();
                    int b = iv_drag_view.getBottom();
                    int l = iv_drag_view.getLeft();
                    int r = iv_drag_view.getRight();
                    //获取到移动后的View的在窗体中的位置
                    int newl = l + dx;
                    int newt = t + dy;
                    int newr = r + dx;
                    int newb = b + dy;
//通过对移动刚结束的View距离手机屏幕四个边框的大小的判断,来避免View被移出屏幕
                    if (newl < 0 || newt < 0 || newr > windowWidth
                            || newb > windowHeight) {
                        break;
                    }
                    //将移动后的View在窗体上重新显示出来
                    iv_drag_view.layout(newl, newt, newr, newb);
                    //立即更新手指第一次触摸屏幕的位置坐标,以便下次继续移动
                    startx = (int) event.getRawX();
                    starty = (int) event.getRawY();
                    Log.i(TAG, "移动");
                    break;
                case MotionEvent.ACTION_UP://手指离开屏幕
                    Log.i(TAG, "松手");
//记录当前imageview在窗体中的位置(左上角的顶点距离屏幕的宽度和高度)
                    Editor editor = sp.edit();
                    int lasty = iv_drag_view.getTop();
                    int lastx = iv_drag_view.getLeft();
                    editor.putInt("lastx", lastx);
                    editor.putInt("lasty", lasty);
                    editor.commit();
                    break;
            }
            //true 会消费调当前的触摸事件,那么后面的移动和离开事件会被响应到
            //false 不会消费当前的触摸事件,那么后面的移动和离开事件都不会被响应到
            return true;
        }
    });
    }
}
```

代码解析:

(1) new Thread() {
        public void run() {
            try {

```
                    Thread.sleep(500);
                    firstclicktime = 0;
                } catch (InterruptedException e) {
                    e.printStackTrace();
                }
            }
        };
    }.start();
```

该线程是用于解决当用户单击一次屏幕后，停留一段时间后双击屏幕的情况。如果去除这块代码，程序的流程是：首先执行 firstclicktime = System.currentTimeMillis()，此时的 firstclicktime>0，当停留 2 秒后执行双击操作，第一次单击时 if (firstclicktime > 0)就为 true，判断设置的阈值时也为 true（时长已经大于 2 秒），此时就已经执行双击水平居中的效果了。但这并不是我们通常所说的双击！如果上面的代码存在，当第一次单击时间超过 0.5 秒时，此时将 firstclicktime = System.currentTimeMillis()中的 firstclicktime 的值设置为 0，就不会出现"伪双击"了。

（2）iv_drag_view.setLayoutParams(params)

这是用于初始化上次 View 被移动的位置。我们在这里并没有使用 View 移动后重新设置 View 在窗体中位置的那句代码：iv_drag_view.layout(newl, newt, newr, newb)，如果使用后面的代码，程序是会出现错误的。

控件被显示到界面上，其实经历了两个过程：

> 计算控件在窗体中的位置和大小；
> 在窗体中移动一个控件。

而 onCreate 方法初始化界面时，是在第一个阶段，该阶段用来测量控件的大小和位置。所以，初始化上次移动后的 View 的显示位置应当放在第一个过程中。

简单介绍一下触摸事件和单击事件的区别。

> 单击事件：是一组动作的集合，包括按下→停留→离开；
> 触摸事件：任何一个屏幕事件都是一个触摸（持续时间可能很长）；
> 触摸事件的监听器对应的事件的处理 onTouch 和单击事件对应的事件的处理 onClick。如果两者共存的话，前者执行的优先级高于后者。而且，如果事件在 onTouch 中没有消费掉，那么在 onClick 中会被消费掉；如果在 onTouch 中已经被消费掉，那么在 onClick 中将不会捕捉到该事件。

测试运行：当双击 View 时，View 并没有居中。原因在于：单击事件在 onTouch 中已经被消费掉了，而在 onClick 中则不会被捕获到。所以，在 onTouch 中不消费此事件，该事件最终会在 onClick 中被消费掉。测试运行后的结果界面如图 3-17 所示。

图 3-17

最后，在来电归属地信息显示的服务中将 View 的对应参数设置进来，ShowCallLocationService.java 中的业务代码如下：

```
package com.guoshisp.mobilesafe.service;
import android.app.Service;
import android.content.Intent;
import android.content.SharedPreferences;
import android.graphics.PixelFormat;
import android.os.IBinder;
import android.telephony.PhoneStateListener;
import android.telephony.TelephonyManager;
import android.view.Gravity;
import android.view.View;
import android.view.WindowManager;
import android.view.WindowManager.LayoutParams;
import android.widget.LinearLayout;
import android.widget.TextView;
import com.guoshisp.mobilesafe.R;
import com.guoshisp.mobilesafe.db.dao.NumberAddressDao;
//在后台监听电话呼入的状态
public class ShowCallLocationService extends Service {
    private TelephonyManager tm;//电话管理器
    private MyPhoneListener listener;//电话状态改变的监听器
    private WindowManager windowManager;//窗体管理器
    private SharedPreferences sp;//用于取出归属地风格显示风格的Item对应的id
    //"半透明","活力橙","卫士蓝","苹果绿","金属灰"
    private static final  int[] bgs = {R.drawable.call_locate_white,R.
                                       drawable.call_locate_orange,
       R.drawable.call_locate_blue,R.drawable.call_locate_green,R.
       drawable.call_locate_gray};
```

```java
@Override
public IBinder onBind(Intent intent) {
    return null;
}
/**
 * 当服务第一次被创建时调用
 */
@Override
public void onCreate() {
    super.onCreate();
    sp =getSharedPreferences("config",MODE_PRIVATE);
    //注册系统的电话状态改变的监听器
    listener = new MyPhoneListener();
    //获取系统的电话管理器
    tm = (TelephonyManager) getSystemService(TELEPHONY_SERVICE);
    //为电话设置一个监听。参数一：监听器；参数二：要监听的电话改变类型（这里监听
      的是通话状态）
    tm.listen(listener, PhoneStateListener.LISTEN_CALL_STATE);
    windowManager = (WindowManager) getSystemService(WINDOW_SERVICE);
}
private class MyPhoneListener extends PhoneStateListener {
    private View view;
    //参数一：手机的状态；参数二：呼叫进来的手机号码
    @Override
    public void onCallStateChanged(int state, String incomingNumber) {
        switch (state) {
        case TelephonyManager.CALL_STATE_RINGING://手机铃声正在响
            //获取呼叫进来号码的地址（查询之前的号码归属地数据库）
            String address = NumberAddressDao.getAddress
                    (incomingNumber);
            //使用系统的吐司来显示归属地信息，但显示的时间较短
            //Toast.makeText(getApplicationContext(), "归属地:"+address,
              1).show();
    //通过布局填充器将一个显示号码归属地的布局转成View,该View是一个吐司
            view = View.inflate(getApplicationContext(), R.layout.
                    show_address,null);
            //获取到显示号码归属地布局的根布局LinearLayout
            LinearLayout ll = (LinearLayout) view.findViewById(R.id.
                    ll_show_address);
            //从sp文件中获取显示归属地风格的Item的id
            int which = sp.getInt("which", 0);
            //设置来电归属地显示的背景图片
            ll.setBackgroundResource(bgs[which]);
            //查找view中的用于显示归属地的TextView
            TextView tv = (TextView) view.findViewById(R.id.tv_show_
                    address);
```

```java
            //将归属地信息设置到TextView
            tv.setText(address);
//获取到与窗体相关的布局的参数（这里用于设置窗体上显示来电归属地的吐司的参数信息）
            final WindowManager.LayoutParams params = new LayoutParams();
            //指定吐司的重心为图形的左上角对应的点
            params.gravity = Gravity.LEFT | Gravity.TOP;
//设置吐司在窗体中的显示位置。获取到吐司离窗体左端的X值、获取到吐司离窗体顶端的Y值
            params.x = sp.getInt("lastx", 0);
            params.y = sp.getInt("lasty", 0);
            //设置窗体布局View的高度
            params.height = WindowManager.LayoutParams.WRAP_CONTENT;
            //设置窗体布局View的宽度
            params.width = WindowManager.LayoutParams.WRAP_CONTENT;
            //窗体View不可以获取焦点、不可以被触摸、保持在屏幕上
            params.flags = WindowManager.LayoutParams.FLAG_NOT_FOCUSABLE
                    | WindowManager.LayoutParams.FLAG_NOT_TOUCHABLE
                    | WindowManager.LayoutParams.FLAG_KEEP_SCREEN_ON;
            //显示在窗体上的style为半透明
            params.format = PixelFormat.TRANSLUCENT;
            //窗体View的类型为吐司
            params.type = WindowManager.LayoutParams.TYPE_TOAST;
            //将吐司挂载在窗体上。窗体服务是一个全局的系统服务，该服务开启后会在后
              台运行。一般情况下，在窗体上一旦挂载一个View并显示后，并不会自动消失
            windowManager.addView(view, params);
            break;
        case TelephonyManager.CALL_STATE_IDLE: //手机的空闲状态
            if(view!=null){
                //将窗体上的吐司移除
                windowManager.removeView(view);
                view = null;
            }
            break;
        case TelephonyManager.CALL_STATE_OFFHOOK://手机接通通话时的状态
            if(view!=null){
                //将窗体上的吐司移除
                windowManager.removeView(view);
                view = null;
            }
            break;
        }
        super.onCallStateChanged(state, incomingNumber);
    }
}
/**
 * 取消电话状态的监听
 */
```

```
    @Override
    public void onDestroy() {
        super.onDestroy();
        tm.listen(listener, PhoneStateListener.LISTEN_NONE);
        listener = null;
    }
}
```

这样，当再次呼入一个号码时，号码的归属地就应该显示在我们之前设置的位置上。

测试运行：开启"来电归属地设置"，模拟一个手机号码的到来，界面效果如图 3-19 所示。

图 3-18

到此，来电归属地信息显示的设置全部完成。

3.6 使用 ExpandableListView 实现常用号码的查询

需求：在高级工具中，有一个"常用号码查询"的条目。当第一次单击该条目时，会执行复制数据库的操作，复制完毕后进入到显示常用号码的界面，该界面中将常用号码分类，当单击条目时就会将对应的所有号码以列表的形式展开，单击号码时，进入到拨号界面，当再次单击该条目时，所有号码会收起来。

思路：实现单击一个 View 对象时展开一组 View 对象的效果需要用到 ExpandableList View，在为该 View 绑定数据时需要设置一个对应的适配器，在适配器中首先需要为每个分组绑定数据，再为分组中的 View 绑定数据。实现拨号时需要为每个分组中的 View 设置一个监听器，在监听器中将号码传入后激活手机系统中的拨号器即可。

说明：ExpandableListView 的一级界面中的条目可以称为"分组"，分组中的子 View 可以称为"子孩子"。

首先需要将常用号码的数据库 commonnum.db 复制到 assets 目录下。下面，我们来看下数据库的表结构中的核心信息：表"classlist"是分组所对应的文字的显示，由图 3-19 可以看出表"classlist"一共分为 8 组，表"table"是分组中的每个子 View 的具体信息，所以一共有 8 个"table"表，表"table1"如图 3-20 所示。需要注意的一点的是，"table"表中"_id"的起始值是 1 而不是 0，在后面需要用到。

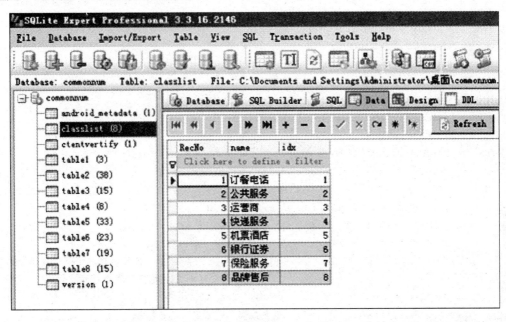

图 3-19

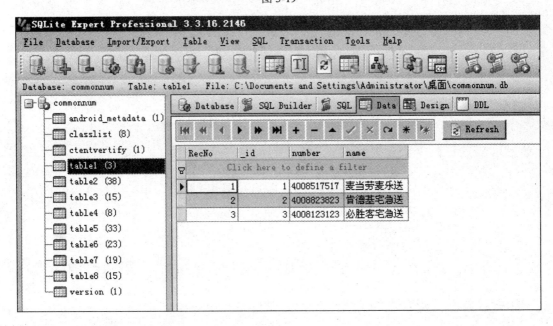

图 3-20

因为，我们要实现的"常用号码查询"功能是在"高级工具"中实现的，所以，AtoolsActivity.java 的业务代码如下：

```java
package com.guoshisp.mobilesafe;
import java.io.File;
import android.app.Activity;
import android.app.ProgressDialog;
import android.content.Intent;
import android.os.Bundle;
import android.os.Handler;
import android.os.Message;
import android.view.View;
import android.view.View.OnClickListener;
import android.widget.TextView;
import android.widget.Toast;
import com.guoshisp.mobilesafe.utils.AssetCopyUtil;
public class AtoolsActivity extends Activity implements OnClickListener {
    protected static final int COPY_SUCCESS = 30;
    protected static final int COPY_FAILED = 31;
    protected static final int COPY_COMMON_NUMBER_SUCCESS = 32;
    private TextView tv_atools_address_query;//当单击该条目时，要执行复制号码
                                              归属地信息的数据库文件
    private TextView tv_atools_common_num;//常用号码
    private ProgressDialog pd;//复制数据库时要显示的进度条
    //复制数据库是一个相对耗时的操作，复制完成后，给主线程发送消息
    private Handler handler = new Handler() {
        public void handleMessage(android.os.Message msg) {
            //无论复制是否成功，都需要关闭进度显示条
            pd.dismiss();
            switch (msg.what) {
            case COPY_SUCCESS:
                //复制数据库成功后，进入号码归属地查询的界面
                loadQueryUI();
                break;
            case COPY_COMMON_NUMBER_SUCCESS:
                //复制数据库成功后，进入常用号码显示的界面
                loadCommNumUI();
                break;
            case COPY_FAILED:
                Toast.makeText(getApplicationContext(), "复制数据失败", 0).
                    show();
                break;
            }
        }
    };
```

```java
@Override
protected void onCreate(Bundle savedInstanceState) {
    super.onCreate(savedInstanceState);
    setContentView(R.layout.atools);//高级工具对应的界面
    tv_atools_address_query = (TextView) findViewById(R.id.tv_atools_
                    address_query);
    tv_atools_address_query.setOnClickListener(this);
    pd = new ProgressDialog(this);
    tv_atools_common_num = (TextView) findViewById(R.id.tv_atools_
                    common_num);
    tv_atools_common_num.setOnClickListener(this);
    //设置进度条显示的风格
    pd.setProgressStyle(ProgressDialog.STYLE_HORIZONTAL);
}
public void onClick(View v) {
    switch (v.getId()) {
    case R.id.tv_atools_address_query://号码归属地查询
//创建出数据库要复制到的系统文件: data\data\包名\files\address.db
        final File file = new File(getFilesDir(), "address.db");
//判断数据库是否存在,如果存在,则直接进入号码归属地的查询界面,否则,执行复制动作
        if (file.exists() && file.length() > 0) {
            //数据库文件复制成功,进入查询号码归属地界面
            loadQueryUI();
        } else {
            //数据库的复制。开始复制时需要开始显示进度条
            pd.show();
            //复制数据库也是一个相对耗时的操作,在子线程中执行该操作
            new Thread() {
                public void run() {
                    AssetCopyUtil asu = new AssetCopyUtil(
                            getApplicationContext());
                    //返回复制成功与否的结果
            boolean result = asu.copyFile("naddress.db", file, pd);
                    if (result) {//复制成功
                        Message msg = Message.obtain();
                        msg.what = COPY_SUCCESS;
                        handler.sendMessage(msg);
                    } else {//复制失败
                        Message msg = Message.obtain();
                        msg.what = COPY_FAILED;
                        handler.sendMessage(msg);
                    }
                };
            }.start();
        }
```

```java
                break;
            case R.id.tv_atools_common_num://公用号码查询
//判读数据库是否已经复制到系统目录（data/data/包名/files/address.db）
                final File commonnumberfile = new File(getFilesDir(),
                        "commonnum.db");
                if (commonnumberfile.exists() && commonnumberfile.length() > 0) {
                    loadCommNumUI();//进入公共号码的显示界面
                } else {
                    //数据库的复制
                    pd.show();
                    //复制数据库是一个相对耗时的工作,我们为其开启一个子线程
                    new Thread() {
                        public void run() {
                            //将数据库复制到手机系统中
                            AssetCopyUtil asu = new AssetCopyUtil(
                                    getApplicationContext());
                            boolean result = asu.copyFile("commonnum.db",
                                    commonnumberfile, pd);
                            if (result) {//复制成功
                                Message msg = Message.obtain();
                                msg.what = COPY_COMMON_NUMBER_SUCCESS;
                                handler.sendMessage(msg);
                            } else {//复制失败
                                Message msg = Message.obtain();
                                msg.what = COPY_FAILED;
                                handler.sendMessage(msg);
                            }
                        };
                    }.start();
                }
                break;
        }
    }
    /**
     * 进入常用号码界面
     */
    private void loadCommNumUI() {
        Intent intent = new Intent(this, CommonNumActivity.class);
        startActivity(intent);
    }
    /**
     * 进入到号码归属地查询界面
     */
    private void loadQueryUI() {
        Intent intent = new Intent(this, NumberQueryActivity.class);
```

```
            startActivity(intent);
        }
}
```

该界面所对应的布局文件 atools.xml 文件内容如下：

```xml
<?xml version="1.0" encoding="utf-8"?>
<LinearLayout xmlns:android="http://schemas.android.com/apk/res/android"
    android:layout_width="fill_parent"
    android:layout_height="fill_parent"
    android:orientation="vertical" >
    <TextView
        android:layout_width="wrap_content"
        android:layout_height="wrap_content"
        android:layout_gravity="center_horizontal"
        android:text="高级工具"
        android:textColor="#66ff00"
        android:textSize="28sp" />
    <View
        android:layout_width="fill_parent"
        android:layout_height="1dip"
        android:layout_marginTop="5dip"
        android:background="@drawable/devide_line" />
    <TextView
         android:layout_marginTop="8dip"
        android:id="@+id/tv_atools_address_query"
        style="@style/text_content_style"
        android:text="号码归属地查询" />
    <View
        android:layout_width="fill_parent"
        android:layout_height="1dip"
        android:layout_marginTop="4dip"
        android:background="@drawable/listview_devider" >
    </View>
    <TextView
         android:layout_marginTop="8dip"
        android:id="@+id/tv_atools_common_num"
        style="@style/text_content_style"
        android:text="常用号码查询" />
    <View
        android:layout_width="fill_parent"
        android:layout_height="1dip"
        android:layout_marginTop="4dip"
        android:background="@drawable/listview_devider" >
    </View>
</LinearLayout>
```

如果常用号码数据库复制成功后，则进入常用号码显示的界面，常用号码显示的 Activity 为 CommonNumActivity，CommonNumActivity.java 业务代码如下：

```java
package com.guoshisp.mobilesafe;
import java.util.HashMap;
import java.util.List;
import java.util.Map;
import android.app.Activity;
import android.content.Intent;
import android.net.Uri;
import android.os.Bundle;
import android.view.View;
import android.view.ViewGroup;
import android.widget.BaseExpandableListAdapter;
import android.widget.ExpandableListView;
import android.widget.ExpandableListView.OnChildClickListener;
import android.widget.TextView;
import com.guoshisp.mobilesafe.db.dao.CommonNumDao;
public class CommonNumActivity extends Activity {
    protected static final String TAG = "CommonNumActivity";
    private ExpandableListView elv_common_num;//可扩展的ListView
    @Override
    protected void onCreate(Bundle savedInstanceState) {
        super.onCreate(savedInstanceState);
        setContentView(R.layout.common_num);
        elv_common_num = (ExpandableListView) findViewById(R.id.elv_
                    common_num);
        elv_common_num.setAdapter(new CommonNumberAdapter());//为
ExpandableListView 设置一个适配器对象，该对象需要是 ExpandableListAdapter
对象的子类
        //为分组中的每个孩子注册一个监听器
        elv_common_num.setOnChildClickListener(new OnChildClickListener() {
            public boolean onChildClick(ExpandableListView parent, View v,
                    int groupPosition, int childPosition, long id) {
                //获取到 TextView 中的电话号码
                TextView tv = (TextView) v;
                String number = tv.getText().toString().split("\n")[1];
                //使用隐式意图来激活手机系统中的拨号器
                Intent intent = new Intent();
                intent.setAction(Intent.ACTION_DIAL);
                intent.setData(Uri.parse("tel:"+number));
                startActivity(intent);
                return false;
            }
```

```java
    });
}
//ExpandableListView的适配器对象，该对象是ExpandableListAdapter对象的子类
private class CommonNumberAdapter extends BaseExpandableListAdapter {
    private List<String> groupNames;
    //将子孩子的所有信息一次性从数据库中获取出来，这样可以避免重复查询数据库内存
        缓存集合。key：分组的位置；value：分组里面所有子孩子的信息
    private Map<Integer, List<String>> childrenCache;
    public CommonNumberAdapter() {
        childrenCache = new HashMap<Integer, List<String>>();
    }
    /**
     * 返回当前列表有多少组
     */
    public int getGroupCount() {
        return CommonNumDao.getGroupCount();
    }
    /**
     * 返回每一组里面有多少个条目
     */
    public int getChildrenCount(int groupPosition) {
        return CommonNumDao.getChildrenCount(groupPosition);
    }
    /**
     * 返回分组所对应的对象。这里用不到，所以返回null
     */
    public Object getGroup(int groupPosition) {
        return null;
    }
    /**
     * 获取分组中的条目对象。这里我们用不到，所以返回null
     */
    public Object getChild(int groupPosition, int childPosition) {
        return null;
    }
    /**
     * 获取分组所对应的id
     */
    public long getGroupId(int groupPosition) {
        return groupPosition;
    }
    /**
     * 获取分组中的条目所对应的id
     */
```

```java
public long getChildId(int groupPosition, int childPosition) {
    return childPosition;
}
/**
 * 是否要为分组中的条目设置一下 id。false 代表不用设置
 */
public boolean hasStableIds() {
    return false;
}
/**
 * 返回每一个分组的 view 对象
 * 参数一：当前分组的 id
 * 参数二：当前分组的 View 是否可扩展
 * 参数三：缓存的 View 对象
 * 参数四：当前分组的父 View 对象
 */
public View getGroupView(int groupPosition, boolean isExpanded,
        View convertView, ViewGroup parent) {
    TextView tv;
    //使用缓存的 View 对象
    if (convertView == null) {
        tv = new TextView(getApplicationContext());
    } else {
        tv = (TextView) convertView;
    }
    tv.setTextSize(28);
    if (groupNames != null) {
        tv.setText("        " + groupNames.get(groupPosition));
    } else {
        groupNames = CommonNumDao.getGroupNames();
        tv.setText("        " + groupNames.get(groupPosition));
    }
    return tv;
}
/**
 * 返回每一个分组某一个位置对应的孩子的 view 对象
 * 参数一：当前分组的 id
 * 参数二：分组中的子孩子的 id
 * 参数三：分组中的子孩子是否是最后一个
 * 参数四：子孩子 View 的缓存对象
 * 参数五：分组中的子孩子所在的父 View 对象
 */
public View getChildView(int groupPosition, int childPosition,
        boolean isLastChild, View convertView, ViewGroup parent) {
    TextView tv;
```

```java
                if (convertView == null) {
                    tv = new TextView(getApplicationContext());
                } else {
                    tv = (TextView) convertView;
                }
                tv.setTextSize(20);
                String result = null;
                if (childrenCache.containsKey(groupPosition)) {
                    result = childrenCache.get(groupPosition).get
                            (childPosition);
                } else {
                    List<String> results = CommonNumDao
                            .getChildNameByPosition(groupPosition);
                    childrenCache.put(groupPosition, results);//把数据放在缓存里面
                    result = results.get(childPosition);
                }
                tv.setText(result);
                return tv;
            }
            /**
             * 返回值如果为true,则表示每个分组的子孩子都可以响应到单击事件,否则,不
                可以响应
             */
            public boolean isChildSelectable(int groupPosition, int
                                        childPosition) {
                return true;
            }
        }
    }
```

其对应的布局文件为common_num.xml文件,ExpandableListView控件在该文件中被使用到,common_num.xml文件中的内容如下:

```xml
<?xml version="1.0" encoding="utf-8"?>
<LinearLayout xmlns:android="http://schemas.android.com/apk/res/android"
    android:layout_width="fill_parent"
    android:layout_height="fill_parent"
    android:orientation="vertical" >
    <TextView
        android:layout_width="wrap_content"
        android:layout_height="wrap_content"
        android:layout_gravity="center_horizontal"
        android:text="常用号码"
        android:textColor="#66ff00"
        android:textSize="28sp" />
    <View
```

```xml
        android:layout_width="fill_parent"
        android:layout_height="1dip"
        android:layout_marginTop="5dip"
        android:background="@drawable/devide_line" />
    <ExpandableListView
        android:id="@+id/elv_common_num"
        android:layout_width="fill_parent"
        android:layout_height="fill_parent" >
    </ExpandableListView>

</LinearLayout>
```

代码解析：

（1）private List<String> groupNames;//存储对应组中的子孩子的详细信息
private Map<Integer, List<String>> childrenCache;

两者在被赋值时，都是从数据库中将对应的数据一次性地取出存入这两个缓存中。这么做的原因在于：一次性将单组的所有信息存入缓存中后，在以后为对应组中的 View 赋值时可以直接从该缓存中取出，而避对数据库的频繁操作。如果每次需要从数据库中取出数据，当用户不断地滑动 ExpandableListView 中的子条目时会有卡屏的感觉，而且后台还会产生大量的 GC 信息。

（2）if (convertView == null) {
　　　　tv = new TextView(getApplicationContext());
　　} else {
　　　　tv = (TextView) convertView;
　　}

这是复用缓存 View 的操作，如果 tv 已经存在，那么下次需要使用到 tv 时，将直接使用上次的 tv，而不用重新创建。这也是对 ExpandableListView 的一种优化——复用缓存。

对数据库进行操作的对象 CommonNumDao 存放在包"com.guoshisp.mobilesafe.db.dao"下，CommonNumDao.java 中的代码如下：

```java
package com.guoshisp.mobilesafe.db.dao;
import java.util.ArrayList;
import java.util.List;
import android.database.Cursor;
import android.database.sqlite.SQLiteDatabase;
public class CommonNumDao {
    /**
     * 返回数据库有多少个分组
     *
     * @return
     */
    public static int getGroupCount() {
```

```java
        int count = 0;
        //所要打开的数据库在手机系统中的位置
        String path = "/data/data/com.guoshisp.mobilesafe/files/
                commonnum.db";
        //打开数据库
        SQLiteDatabase db = SQLiteDatabase.openDatabase(path, null,
                SQLiteDatabase.OPEN_READONLY);
        if (db.isOpen()) {
            //从classlist表中查询出有多少组公用号码
            Cursor cursor = db.rawQuery("select * from classlist", null);
            count = cursor.getCount();
            //使用完数据库后需要关闭
            cursor.close();
            db.close();
        }
        return count;
    }
    /**
     * 获取所有的分组集合信息（也即每个分组的名字）
     *
     * @return
     */
    public static List<String> getGroupNames() {
        //用于存放各个分组的名字信息
        List<String> groupNames = new ArrayList<String>();
        String path = "/data/data/com.guoshisp.mobilesafe/files/
                commonnum.db";
        SQLiteDatabase db = SQLiteDatabase.openDatabase(path, null,
                SQLiteDatabase.OPEN_READONLY);
        if (db.isOpen()) {
            //获取各个分组的名字的结果集（游标）
            Cursor cursor = db.rawQuery("select name from classlist", null);
            //遍历出每个分组对应的名字
            while (cursor.moveToNext()) {
                String groupName = cursor.getString(0);
                groupNames.add(groupName);
                groupName = null;
            }
            cursor.close();
            db.close();
        }
        return groupNames;
    }
    /**
     * 通过单击的分组对应的id来获取某该分组名称
     *
```

```java
 * @param groupPosition
 * @return
 */
public static String getGroupNameByPosition(int groupPosition) {
    String name = null;
    String path = "/data/data/com.guoshisp.mobilesafe/files/
            commonnum.db";
    //因为classlist表中的name的id是从1开始的,而ExpandableListView中
        的id是从0开始的
    int newposition = groupPosition + 1;
    SQLiteDatabase db = SQLiteDatabase.openDatabase(path, null,
            SQLiteDatabase.OPEN_READONLY);
    if (db.isOpen()) {
        //通过id的查询,来获取该id多对应的name
        Cursor cursor = db.rawQuery(
                "select name from classlist where idx=?",
                new String[] { newposition + "" });
        if (cursor.moveToFirst()) {//cursor指针的默认位置是在第一条数据上面
                                   的,所以,想获取数据,指针必须往下移动
            //因为一个id只是对应一个name,所以只需要获取到第一个即可
            name = cursor.getString(0);
        }
        cursor.close();
        db.close();
    }
    return name;
}
/**
 * 获取对应的分组里面有多少个子孩子(也即每个分组里面有多少条号码)
 * 而每个孩子都各自对应一张表,所以在查询子孩子的信息时需要确定是查询哪张孩子所对
    应的表:table+position
 * @param groupPosition
 * @return
 */
public static int getChildrenCount(int groupPosition) {
    int count = 0;
    String path = "/data/data/com.guoshisp.mobilesafe/files/
            commonnum.db";
    //打开数据库
    SQLiteDatabase db = SQLiteDatabase.openDatabase(path, null,
            SQLiteDatabase.OPEN_READONLY);
    //因为groupPosition的起始值是从开始的,而table表中的_id是从1开始的
    int newposition = groupPosition + 1;
    String sql = "select * from table" + newposition;
    if (db.isOpen()) {
```

```java
            Cursor cursor = db.rawQuery(sql, null);
            //获取到查询结果的所有列数（也相当于有多少条号码）
            count = cursor.getCount();
            cursor.close();
            db.close();
        }
        return count;
    }
    /**
     * 获取对应位置的子孩子的信息
     * 而每个孩子都各自对应一张表,所以在查询子孩子的信息时需要确定是查询哪张孩子所对
       应的表: table+position
     */
    public static String getChildNameByPosition(int groupPosition,
            int childPosition) {
        String result = null;
        String path = "/data/data/com.guoshisp.mobilesafe/files/
                    commonnum.db";
        SQLiteDatabase db = SQLiteDatabase.openDatabase(path, null,
                SQLiteDatabase.OPEN_READONLY);
        int newGroupPosition = groupPosition + 1;
        int newChildPosition = childPosition + 1;
        //查询子孩子的 name 和 number
        String sql = "select name,number from table" + newGroupPosition
                + " where _id=?";
        if (db.isOpen()) {
            Cursor cursor = db.rawQuery(sql, new String[] { newChildPosition
                    + "" });
            if (cursor.moveToFirst()) {
        //因为查询的是 name 和 number,且 name 在前,number 在后,所以是两条信息
                String name = cursor.getString(0);
                String number = cursor.getString(1);
                result = name + "\n" + number;
            }
            cursor.close();
            db.close();
        }
        return result;
    }
    /**
     * 获取每一个分组所有的子孩子的信息（name 和 number）
     * 而每个孩子都各自对应一张表,所以在查询子孩子的信息时需要确定是查询哪张孩子所对
       应的表: table+position
     */
    public static List<String> getChildNameByPosition(int groupPosition) {
        String result = null;
```

```
        List<String> results = new ArrayList<String>();
        String path = "/data/data/com.guoshisp.mobilesafe/files/
                   commonnum.db";
        SQLiteDatabase db = SQLiteDatabase.openDatabase(path, null,
                SQLiteDatabase.OPEN_READONLY);
        int newGroupPosition = groupPosition + 1;
        //查询子孩子的 name 和 number
        String sql = "select name,number from table" + newGroupPosition;
        if (db.isOpen()) {
            Cursor cursor = db.rawQuery(sql, null);
            while (cursor.moveToNext()) {
        //因为查询的是 name 和 number，且 name 在前，number 在后，所以是两条信息
                String name = cursor.getString(0);
                String number = cursor.getString(1);
                result = name + "\n" + number;
                results.add(result);
                result = null;
            }
            cursor.close();
            db.close();
        }
        return results;
    }
}
```

测试运行：为 CommonNumActivity 在清单文件中做对应的配置信息后，第一次单击高级工具中的"常用号码查询"条目时的界面如图 3-21 所示（复制公用号码数据库）；之后进入到"常用号码"显示界面，如图 3-22 所示；单击"快递服务"条目后的界面效果如图 3-23 所示；单击其中的"顺丰快递"后的界面如图 3-24 所示。

图 3-21　　　　　　　　　　　图 3-22

第 3 章 高级工具模块的设计

图 3-23　　　　　　　　　　　图 3-24

3.7　程序锁的设计和 UI

需求：在高级工具中有一个"程序锁"条目，当单击该条目后会进入一个新的界面，在该界面中会列出当前手机中的所有应用程序。单击某个应用程序时会实现对应该应用程序的锁定与解锁，当某个应用程序被锁定后，下次再进入该应用程序时就需要输入正确的密码后才可以进入。同时，我们在设置中心对程序锁服务是否开启做一个设置，只有开启后，程序锁才会生效。这样就实现了对用户隐私的保护。

思路：采用服务（看门狗服务）实时获取当前手机中位于栈顶的 Activity 应用程序所对应的包名（位于栈顶，说明该 Activity 处于显示的状态）→判断该包名是否是要被锁定的包名→如果是，则弹出一个输入密码的对话框，当输入正确的密码后，服务临时停止对该应用的保护，临时将该应用程序的包名存入一个缓存中（定义一个集合）。如果设置中心开启程序锁功能，则开启该后台服务（看门狗服务）；如果设置中心没有开启程序锁功能，则停止该服务。为保证数据的同步，我们自定义了一个内容提供者（没有使用使用内容观察者的原因在于：让读者更深一步地了解内容提供者的使用，同时，这也让程序锁功能的实现同时使用了四大组件）。

当一个应用程序被保护后，我们将其对应的包名存入数据库中；当一个被保护的程序解锁后，我们将其对应的包名从数据库中移除。

程序锁的 UI 界面如图 3-25 所示。

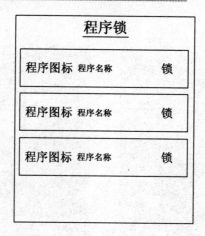

图 3-25

3.7.1 程序锁的实现

首先,在"高级工具"中添加"程序锁"条目,并为该条目设置单击事件。此时,AtoolsActivity.java 中的最终业务代码如下:

```
package com.guoshisp.mobilesafe;
import java.io.File;
import android.app.Activity;
import android.app.ProgressDialog;
import android.content.Intent;
import android.os.Bundle;
import android.os.Handler;
import android.os.Message;
import android.view.View;
import android.view.View.OnClickListener;
import android.widget.TextView;
import android.widget.Toast;
import com.guoshisp.mobilesafe.utils.AssetCopyUtil;
public class AtoolsActivity extends Activity implements OnClickListener {
    protected static final int COPY_SUCCESS = 30;
    protected static final int COPY_FAILED = 31;
    protected static final int COPY_COMMON_NUMBER_SUCCESS = 32;
    private TextView tv_atools_address_query;//当单击该条目时,要执行复制号码
                                             归属地信息的数据库文件
    private TextView tv_atools_common_num;//常用号码
    private TextView tv_atools_applock;//程序锁
    private ProgressDialog pd;//复制数据库时要显示的进度条
    //复制数据库是一个相对耗时的操作,复制完成后,给主线程发送消息
    private Handler handler = new Handler() {
        public void handleMessage(android.os.Message msg) {
            //无论复制是否成功,都需要关闭进度显示条
```

```java
            pd.dismiss();
            switch (msg.what) {
            case COPY_SUCCESS:
                //复制数据库成功后,进入号码归属地查询的界面
                loadQueryUI();
                break;
            case COPY_COMMON_NUMBER_SUCCESS:
                //复制数据库成功后,进入常用号码显示的界面
                loadCommNumUI();
                break;
            case COPY_FAILED:
                Toast.makeText(getApplicationContext(), "复制数据失败", 0).
                    show();
                break;
            }
        };
    };
    @Override
    protected void onCreate(Bundle savedInstanceState) {
        super.onCreate(savedInstanceState);
        setContentView(R.layout.atools);//高级工具对应的界面
        tv_atools_address_query = (TextView) findViewById(R.id.tv_atools_
                        address_query);
        tv_atools_address_query.setOnClickListener(this);
        pd = new ProgressDialog(this);
        tv_atools_common_num = (TextView) findViewById(R.id.tv_atools_
                        common_num);
        tv_atools_common_num.setOnClickListener(this);
        tv_atools_applock = (TextView)findViewById(R.id.tv_atools_applock);
        tv_atools_applock.setOnClickListener(this);
        //设置进度条显示的风格
        pd.setProgressStyle(ProgressDialog.STYLE_HORIZONTAL);
    }
    public void onClick(View v) {
        switch (v.getId()) {
        case R.id.tv_atools_address_query://号码归属地查询
            //创建出数据库要复制到的系统文件:data\data\包名\files\address.db
            final File file = new File(getFilesDir(), "address.db");
//判断数据库是否存在,如果存在,则直接进入号码归属地的查询界面,否则,执行复制动作
            if (file.exists() && file.length() > 0) {
                //数据库文件复制成功,进入查询号码归属地界面
                loadQueryUI();
            } else {
                //数据库的复制,开始复制时需要开始显示进度条
                pd.show();
                //复制数据库也是一个相对耗时的操作,在子线程中执行该操作
```

```java
            new Thread() {
                public void run() {
                    AssetCopyUtil asu = new AssetCopyUtil(
                            getApplicationContext());
                    //返回复制成功与否的结果
            boolean result = asu.copyFile("naddress.db", file, pd);
                    if (result) {//复制成功
                        Message msg = Message.obtain();
                        msg.what = COPY_SUCCESS;
                        handler.sendMessage(msg);
                    } else {//复制失败
                        Message msg = Message.obtain();
                        msg.what = COPY_FAILED;
                        handler.sendMessage(msg);
                    }
                };
            }.start();
        }
        break;
case R.id.tv_atools_common_num://公用号码查询
    //判读数据库是否已经复制到系统目录（data/data/包名/files/address.db）
    final File commonnumberfile = new File(getFilesDir(),
            "commonnum.db");
    if (commonnumberfile.exists() && commonnumberfile.length() > 0) {
        loadCommNumUI();//进入公共号码的显示界面
    } else {
        //数据库的复制
        pd.show();
        //复制数据库是一个相对耗时的工作，我们为其开启一个子线程
        new Thread() {
            public void run() {
                //将数据库复制到手机系统中
                AssetCopyUtil asu = new AssetCopyUtil(
                        getApplicationContext());
                boolean result = asu.copyFile("commonnum.db",
                        commonnumberfile, pd);
                if (result) {//复制成功
                    Message msg = Message.obtain();
                    msg.what = COPY_COMMON_NUMBER_SUCCESS;
                    handler.sendMessage(msg);
                } else {//复制失败
                    Message msg = Message.obtain();
                    msg.what = COPY_FAILED;
                    handler.sendMessage(msg);
                }
            };
```

```java
                }.start();
            }
            break;
        case R.id.tv_atools_applock://程序锁
            Intent applockIntent = new Intent(this,AppLockActivity.class);
            startActivity(applockIntent);
            break;
        }
    }
    /**
     * 进入常用号码界面
     */
    private void loadCommNumUI() {
        Intent intent = new Intent(this, CommonNumActivity.class);
        startActivity(intent);
    }
    /**
     * 进入到号码归属地查询界面
     */
    private void loadQueryUI() {
        Intent intent = new Intent(this, NumberQueryActivity.class);
        startActivity(intent);
    }
}
```

对应的最终的 atools.xml 文件如下：

```xml
<?xml version="1.0" encoding="utf-8"?>
<LinearLayout xmlns:android="http://schemas.android.com/apk/res/android"
    android:layout_width="fill_parent"
    android:layout_height="fill_parent"
    android:orientation="vertical" >
    <TextView
        android:layout_width="wrap_content"
        android:layout_height="wrap_content"
        android:layout_gravity="center_horizontal"
        android:text="高级工具"
        android:textColor="#66ff00"
        android:textSize="28sp" />
    <View
        android:layout_width="fill_parent"
        android:layout_height="1dip"
        android:layout_marginTop="5dip"
        android:background="@drawable/devide_line" />
    <TextView
        android:id="@+id/tv_atools_address_query"
        style="@style/text_content_style"
```

```xml
        android:layout_marginTop="8dip"
        android:text="号码归属地查询" />
    <View
        android:layout_width="fill_parent"
        android:layout_height="1dip"
        android:layout_marginTop="4dip"
        android:background="@drawable/listview_devider" >
    </View>
    <TextView
        android:id="@+id/tv_atools_common_num"
        style="@style/text_content_style"
        android:layout_marginTop="8dip"
        android:text="常用号码 查询" />
    <View
        android:layout_width="fill_parent"
        android:layout_height="1dip"
        android:layout_marginTop="4dip"
        android:background="@drawable/listview_devider" >
    </View>
    <TextView
        android:id="@+id/tv_atools_applock"
        style="@style/text_content_style"
        android:layout_marginTop="8dip"
        android:text="程 序 锁" />
    <View
        android:layout_width="fill_parent"
        android:layout_height="1dip"
        android:layout_marginTop="4dip"
        android:background="@drawable/listview_devider" >
    </View>
</LinearLayout>
```

程序锁的主界面对应的Activity为AppLockActivity，AppLockActivity.java的业务代码如下：

```java
package com.guoshisp.mobilesafe;
import java.util.List;
import android.app.Activity;
import android.os.Bundle;
import android.os.Handler;
import android.view.View;
import android.view.ViewGroup;
import android.view.animation.Animation;
import android.view.animation.TranslateAnimation;
import android.widget.AdapterView;
import android.widget.AdapterView.OnItemClickListener;
import android.widget.BaseAdapter;
import android.widget.ImageView;
```

```java
import android.widget.LinearLayout;
import android.widget.ListView;
import android.widget.TextView;
import com.guoshisp.mobilesafe.db.dao.AppLockDao;
import com.guoshisp.mobilesafe.domain.AppInfo;
import com.guoshisp.mobilesafe.engine.AppInfoProvider;
public class AppLockActivity extends Activity {
    //展示手机中的所有应用
    private ListView lv_applock;
    //ProgressBar 和 TextView 对应的父控件，用于控制 ProgressBar 和 TextView 的显示
    private LinearLayout ll_loading;
    //获取手机中已安装的应用程序
    private AppInfoProvider provider;
    //存放当前手机上所有应用程序的信息
    private List<AppInfo> appinfos;
    //操作存放已锁定的应用程序的数据库
    private AppLockDao dao;
    //存放所有已经被锁定的应用程序的包名信息
    private List<String> lockedPacknames;
    //处理子线程中获取到的当前手机中所有应用程序
    private Handler handler = new Handler(){
        public void handleMessage(android.os.Message msg) {
            ll_loading.setVisibility(View.INVISIBLE);
            //为 ListView 适配数据
            lv_applock.setAdapter(new AppLockAdapter());
        };
    };
    @Override
    protected void onCreate(Bundle savedInstanceState) {
        setContentView(R.layout.app_lock);
        super.onCreate(savedInstanceState);
        provider = new AppInfoProvider(this);
        lv_applock = (ListView) findViewById(R.id.lv_applock);
        ll_loading = (LinearLayout) findViewById(R.id.ll_applock_loading);
        dao =new AppLockDao(this);
        //从数据库中获取到所有被锁定的应用程序包名
        lockedPacknames = dao.findAll();
        //正在从数据库中获取数据时，应该显示 ProgressBar 和 TextView 对应的 "正在加
          载…" 字样
        ll_loading.setVisibility(View.VISIBLE);
        //开启一个子线程获取手机中所有应用程序的信息
        new Thread(){
            public void run() {
                appinfos = provider.getInstalledApps();
                //向主线程中发送一个空消息，通知主线程更新数据
```

```java
                    handler.sendEmptyMessage(0);
                };
            }.start();
            //为ListView中的Item设置单击事件的监听器
            lv_applock.setOnItemClickListener(new OnItemClickListener() {
                public void onItemClick(AdapterView<?> parent, View view,
                        int position, long id) {
                    //获取当前Item的对象
                    AppInfo appinfo = (AppInfo) lv_applock.getItemAtPosition
                                (position);
                    //获取到当前Item对象的包名信息
                    String packname = appinfo.getPackname();
                    //查找到Item对应的锁控件（ImageView）
                    ImageView iv = (ImageView) view.findViewById(R.id.iv_
                                applock_status);
                    //设置一个左右移动的动画
                    TranslateAnimation ta = new TranslateAnimation(Animation.
                                RELATIVE_TO_SELF, 0, Animation.
                                RELATIVE_TO_SELF, 0.2f, Animation.
                                RELATIVE_TO_SELF, 0, Animation.
                                RELATIVE_TO_SELF, 0);
                    //设置动画播放的时长（毫秒）
                    ta.setDuration(200);
                    //判断当前的Item是否处于锁定状态，如果是，则应该解锁，否则应该加锁
                    if(lockedPacknames.contains(packname)){//锁定状态
                        dao.delete(packname);
                        //设置为未锁定状态
                        iv.setImageResource(R.drawable.unlock);
            //将当前应用程序的包名从集合（存放已锁定应用程序的包名）中移除，以便界面的刷新
                        lockedPacknames.remove(packname);
                    }else{//未锁定状态
                        //将包名添加到数据库中
                        dao.add(packname);
                        //设置为锁定状态
                        iv.setImageResource(R.drawable.lock);
            //将当前应用程序的包名添加到集合（存放已锁定应用程序的包名）中，以便界面的刷新
                        lockedPacknames.add(packname);
                    }
                    //为当前的Item播放动画
                    view.startAnimation(ta);
                }
            });
        }
        //自定义适配器对象
        private class AppLockAdapter extends BaseAdapter{
```

```java
    public int getCount() {
        return appinfos.size();
    }
    public Object getItem(int position) {
        return appinfos.get(position);
    }
    public long getItemId(int position) {
        return position;
    }
    public View getView(int position, View convertView, ViewGroup
                    parent) {
        View view;
        ViewHolder holder;
        //复用历史缓存的View对象
        if(convertView==null){
            view = View.inflate(getApplicationContext(),R.layout.app_
                lock_item, null);
            holder = new ViewHolder();
            holder.iv_icon = (ImageView)view.findViewById(R.id.iv_
                        applock_icon);
            holder.iv_status = (ImageView)view.findViewById(R.id.iv_
                        applock_status);
            holder.tv_name = (TextView)view.findViewById(R.id.tv_
                        applock_appname);
            view.setTag(holder);
        }else{//为View做一个标记，以便复用
            view = convertView;
            holder = (ViewHolder) view.getTag();
        }
        //获取到当前应用程序对象
        AppInfo appInfo = appinfos.get(position);
        holder.iv_icon.setImageDrawable(appInfo.getAppicon());
        holder.tv_name.setText(appInfo.getAppname());
    //查看被当前的Item是否是被绑定的应用，以此来为Item设置对应的锁（锁定或未锁定）
        if(lockedPacknames.contains(appInfo.getPackname())){
            holder.iv_status.setImageResource(R.drawable.lock);
        }else{
            holder.iv_status.setImageResource(R.drawable.unlock);
        }
        return view;
    }
}
//View对应的View对象只会在堆内存中存在一份，所有的Item都公用该View
public static class ViewHolder{
    ImageView iv_icon;
```

```
        ImageView iv_status;
        TextView tv_name;
    }
}
```

AppLockActivity 中所对应的布局文件 app_lock.xml 如下：

```xml
<?xml version="1.0" encoding="utf-8"?>
<LinearLayout xmlns:android="http://schemas.android.com/apk/res/android"
    android:layout_width="fill_parent"
    android:layout_height="fill_parent"
    android:orientation="vertical" >
    <TextView
        android:layout_width="wrap_content"
        android:layout_height="wrap_content"
        android:layout_gravity="center_horizontal"
        android:text="程序锁"
        android:textColor="#66ff00"
        android:textSize="28sp" />
    <View
        android:layout_width="fill_parent"
        android:layout_height="1dip"
        android:layout_marginTop="5dip"
        android:background="@drawable/devide_line" />
    <RelativeLayout
        android:layout_width="fill_parent"
        android:layout_height="fill_parent" >
        <LinearLayout
            android:id="@+id/ll_applock_loading"
            android:layout_width="fill_parent"
            android:layout_height="fill_parent"
            android:gravity="center"
            android:orientation="vertical"
            android:visibility="invisible" >
            <ProgressBar
                android:layout_width="wrap_content"
                android:layout_height="wrap_content" />
            <TextView
                android:layout_width="wrap_content"
                android:layout_height="wrap_content"
                android:text="正在加载程序信息…" />
        </LinearLayout>
        <ListView
            android:fastScrollEnabled="true"
            android:id="@+id/lv_applock"
            android:layout_width="fill_parent"
            android:layout_height="fill_parent" >
```

```
        </ListView>
    </RelativeLayout>
</LinearLayout>
```

该文件中的 ListView 有一个属性：android:fastScrollEnabled="true"表示支持 ListView 的快速滚动（在滚动时会出现一个供快速滚动的按钮）。

ListView 中的 Item 对应的布局文件 app_lock_item.xml 如下：

```
<?xml version="1.0" encoding="utf-8"?>
<RelativeLayout xmlns:android="http://schemas.android.com/apk/res/android"
    android:layout_width="match_parent"
    android:gravity="center_vertical"
    android:layout_height="wrap_content" >
    <ImageView
        android:id="@+id/iv_applock_icon"
        android:layout_width="60dip"
        android:layout_height="60dip"
        android:src="@drawable/ic_launcher" />
    <TextView
        android:layout_marginTop="5dip"
        android:singleLine="true"
        android:ellipsize="end"
        android:id="@+id/tv_applock_appname"
        android:layout_width="wrap_content"
        android:layout_height="wrap_content"
        android:layout_toRightOf="@id/iv_applock_icon"
        android:text="程序名称"
        android:textSize="24sp" />
     <ImageView
         android:src="@drawable/unlock"
         android:id="@+id/iv_applock_status"
         android:layout_width="48dip"
         android:layout_height="48dip"
          android:layout_alignParentRight="true"
         />
</RelativeLayout>
```

代码解析：

（1）appinfos = provider.getInstalledApps()

用于获取当前手机中的所有应用程序，provider 对象存放在包"com.guoshisp.mobilesafe.engine"下，AppInfoProvider.java 代码如下：

```
package com.guoshisp.mobilesafe.engine;
import java.util.ArrayList;
import java.util.List;
import android.content.Context;
```

```java
import android.content.pm.ApplicationInfo;
import android.content.pm.PackageInfo;
import android.content.pm.PackageManager;
import com.guoshisp.mobilesafe.domain.AppInfo;
public class AppInfoProvider {
    private PackageManager pm;
    public AppInfoProvider(Context context) {
        pm = context.getPackageManager();
    }
    /**
     * 获取所有安装程序信息
     * @return
     */
    public List<AppInfo> getInstalledApps(){//返回所有的安装的程序列表信息
    //其中,参数PackageManager.GET_UNINSTALLED_PACKAGES 表示包括哪些被卸载的
    //但是没有清除数据的应用
        List<PackageInfo> packageinfos = pm.getInstalledPackages
            (PackageManager. GET_UNINSTALLED_PACKAGES);
        List<AppInfo> appinfos = new ArrayList<AppInfo>();
        for(PackageInfo info : packageinfos){
            AppInfo appinfo = new AppInfo();
            //应用程序的包名
            appinfo.setPackname(info.packageName);
            //应用程序的版本号
            appinfo.setVersion(info.versionName);
            //应用程序的图标 info.applicationInfo.loadIcon(pm);
            appinfo.setAppicon(info.applicationInfo.loadIcon(pm));
            //应用程序的名称 info.applicationInfo.loadLabel(pm);
appinfo.setAppname(info.applicationInfo.loadLabel(pm).toString());
            //过滤出第三方(非系统)应用程序的名称
            appinfo.setUserapp(filterApp(info.applicationInfo));
            appinfos.add(appinfo);
            appinfo = null;
        }
        return appinfos;
    }
    /**
     * 第三方应用程序的过滤器
     * @param info
     * @return true 三方应用
     *    false 系统应用
     */
    public boolean filterApp(ApplicationInfo info) {
    //当前应用程序的标记与系统应用程序的标记
        if ((info.flags & ApplicationInfo.FLAG_UPDATED_SYSTEM_APP) != 0) {
            return true;
```

```
        } else if ((info.flags & ApplicationInfo.FLAG_SYSTEM) == 0) {
            return true;
        }
        return false;
    }
}
```

代码解析：

AppInfo appinfo = new AppInfo()用于封装每个应用程序的具体信息，AppInfo 存放在包"com.guoshisp.mobilesafe.domain"下，AppInfo.java 代码如下：

```
package com.guoshisp.mobilesafe.domain;
import android.graphics.drawable.Drawable;
public class AppInfo {
    //应用包名
    private String packname;
    //应用版本号
    private String version;
    //应用名称
    private String appname;
    //应用图标
    private Drawable appicon;
    //该应用是否属于用户程序
    private boolean userapp;
    public boolean isUserapp() {
        return userapp;
    }
    public void setUserapp(boolean userapp) {
        this.userapp = userapp;
    }
    public String getPackname() {
        return packname;
    }
    public void setPackname(String packname) {
        this.packname = packname;
    }
    public String getVersion() {
        return version;
    }
    public void setVersion(String version) {
        this.version = version;
    }
    public String getAppname() {
        return appname;
    }
    public void setAppname(String appname) {
```

```java
        this.appname = appname;
    }
    public Drawable getAppicon() {
        return appicon;
    }
    public void setAppicon(Drawable appicon) {
        this.appicon = appicon;
    }
}
```

（2）dao =new AppLockDao(this)获取到用于操作程序锁数据库的实例对象。AppLockDao 存放在包"com.guoshisp.mobilesafe.db.dao"下，AppLockDao.java 代码如下：

```java
package com.guoshisp.mobilesafe.db.dao;
import java.util.ArrayList;
import java.util.List;
import android.content.Context;
import android.database.Cursor;
import android.database.sqlite.SQLiteDatabase;
import com.guoshisp.mobilesafe.db.AppLockDBOpenHelper;
public class AppLockDao {
    private AppLockDBOpenHelper helper;
    public AppLockDao(Context context) {
        helper = new AppLockDBOpenHelper(context);
    }
    /**
     * 查找一条锁定程序的包名
     * return true 代表查找到该包名，false 代表没有查找到该包名
     */
    public boolean find(String packname) {
        boolean result = false;
        //打开数据库
        SQLiteDatabase db = helper.getReadableDatabase();
        if (db.isOpen()) {
            //执行查询SQL 语句，返回一个结果集
            Cursor cursor = db.rawQuery(
                    "select * from applock where packname =?",
                    new String[] { packname });
            if (cursor.moveToFirst()) {
                result = true;
            }
            //关闭数据库
            cursor.close();
            db.close();
        }
        return result;
```

```java
    }
    /**
     * 添加一条锁定的程序的包名
     */
    public boolean add(String packname) {
        //首先查询一个数据库中是否存在该条数据，防止重复添加
        if (find(packname))
            return false;
        SQLiteDatabase db = helper.getWritableDatabase();
        if (db.isOpen()) {
            //执行添加的SQL语句
            db.execSQL("insert into applock (packname) values (?)",
                    new Object[] { packname });
            db.close();
        }
        return find(packname);
    }
    /**
     * 删除一条包名
     */
    public void delete(String packname) {
        SQLiteDatabase db = helper.getWritableDatabase();
        if (db.isOpen()) {
            //执行删除的SQL语句
            db.execSQL("delete from applock where packname=?",
                    new Object[] { packname });
            db.close();
        }
    }
    /**
     * 查找全部被锁定的应用包名
     *
     * @return
     */
    public List<String> findAll() {
        List<String> packnames = new ArrayList<String>();
        SQLiteDatabase db = helper.getReadableDatabase();
        if (db.isOpen()) {
            Cursor cursor = db.rawQuery("select packname from applock",
                    null);
            while (cursor.moveToNext()) {
                packnames.add(cursor.getString(0));
            }
            cursor.close();
            db.close();
```

```
            }
            return packnames;
        }
    }
```

以上代码实现了程序锁的 UI 展示，以及对 Item 单击事件的处理。当单击一个处于未锁定状态的 Item 时，表示要将该程序锁定，此时会将其对应的包名添加到程序锁的数据库中，同时也将包名添加到当前的集合缓存中（实现界面信息的同步）；当单击一个处于锁定状态的 Item 时，表示要将该程序解锁，此时会将其对应的包名从程序锁的数据库中删除，同时也将包名从当前的集合缓存中移除（实现界面信息的同步）。测试结果如图 3-26 所示，当单击第一个 Item 后，界面如图 3-27 所示。

看门狗服务是否开启，则可以通过在"设置中心"中添加一个"程序锁设置"条目来实现。如果该条目中的 Checkbox 处于勾选状态，看门狗服务将开启；如果该条目中的 Checkbox 处于未勾选状态，看门狗服务将不会开启。设置中心对应的 SettingCenterActivity.java 的最终业务代码及设置中心对应的最终布局文件 Setting_center.xml 代码见在线资源包中代码文本部分 3.7.1.doc。

图 3-26

图 3-27

设置中心中的看门狗服务 WatchDogService1 存放在包 "com.guoshisp.mobilesafe.service" 下，WatchDogService1.java 如下：

```
package com.guoshisp.mobilesafe.service;
import java.util.List;
import android.app.ActivityManager;
import android.app.ActivityManager.RunningTaskInfo;
import android.app.Service;
import android.content.Intent;
import android.os.IBinder;
import android.util.Log;
```

```java
import com.guoshisp.mobilesafe.EnterPwdActivity;
import com.guoshisp.mobilesafe.db.dao.AppLockDao;
public class WatchDogService1 extends Service {
    protected static final String TAG = "WatchDogService";
    //是否要停止看门狗服务。true表示继续运行，false表示停止运行
    boolean flag;
    //要进入一个已被锁定的应用程序前，需要输入正确的密码后才可以进入，这是一个用于激
      活输入密码的界面
    private Intent pwdintent;
    //将所有已被锁定的应用程序的包名存放在该集合缓存中
    private List<String> lockPacknames;
    //操作数据库的对象
    private AppLockDao dao;
    @Override
    public IBinder onBind(Intent intent) {
        //TODO Auto-generated method stub
        return null;
    }
    @Override
    public void onCreate() {
        super.onCreate();
        dao = new AppLockDao(this);
        //将看门狗服务的标记设置为true，让其一直在后台运行
        flag = true;
        //从程序锁对应的数据库中取出所有应用程序的包名
        lockPacknames = dao.findAll();
        pwdintent = new Intent(this,EnterPwdActivity.class);
        //因为服务本身没有任务栈，如果要开启一个需要在任务栈中运行的Activity，需要
          为该Activity创建一个任务栈
        pwdintent.setFlags(Intent.FLAG_ACTIVITY_NEW_TASK);
        //开启一个线程不断运行看门狗服务
        new Thread() {
            public void run() {
                //设置一个死循环，如果为true，则一直运行
                while (flag){
    //获取一个Activity的管理器，ActivityManager可以动态地观察到当前存在哪些进程
                    ActivityManager am = (ActivityManager) getSystemService
                                    (ACTIVITY_SERVICE);
                    //获取到当前正在栈顶运行的Activity
                    RunningTaskInfo taskinfo = am.getRunningTasks(1).get(0);
                    //获取到当前任务栈顶程序所对应的包名
                    String packname = taskinfo.topActivity.getPackageName();
                    Log.i(TAG,packname);
                    //将任务栈顶的程序的包名信息存入意图中（以键值对的形式存入，可
                      以在被激活的Activity中通过getIntent()来获取该意图，然后
                      再获取意图对象中的数据）
```

```
            pwdintent.putExtra("packname", packname);
            //判断运行在栈顶的程序所对应的包名是否是已锁定的应用程序
            if(lockPacknames.contains(packname)){
            //发现当前应用程序为已锁定的应用程序，需要进入输入密码的界面
                startActivity(pwdintent);
            }
            try {
                //看门狗服务非常耗电，这里用于让该服务暂停200毫秒
                Thread.sleep(200);
            } catch (InterruptedException e) {
                e.printStackTrace();
            }
            }
        };
    }.start();
}
//当服务被停止时，应停止看门狗
@Override
public void onDestroy() {
    flag = false;
    super.onDestroy();
}
}
```

代码解析：

（1）RunningTaskInfo taskinfo = am.getRunningTasks(1).get(0)

获取到任务栈中栈顶的 Activity 的信息。am.getRunningTasks(1)中的参数表示的是要获取到多少个任务栈信息，然后返回一个集合。如果所传入的参数小于实际运行在栈中的信息条数，那么将返回最近运行的条目；如果传入的参数大于任务栈中的实际信息数目，那么将返回任务栈中的所有信息。执行该代码需要在清单文件中配置获取任务栈的权限信息：<uses-permission android:name="android.permission.GET_TASKS" />。

（2）Log.i(TAG,packname)

打印当前运行在栈顶的 Activity 的包名。因为该代码位于一个死循环中，如果这个看门狗一直运行，当切换界面时，会打印出当前界面所对应的包名信息。所以，只要栈顶的 Activity 发生变化，打印的信息也会发生相对应的变化，这就是看门狗服务的核心原理。

（3）startActivity(pwdintent)

如果当前处于栈顶的 Activity 的包名属于锁定状态，就会开启一个新的 Activity，该 Activity 是一个输入密码的界面。

```
package com.guoshisp.mobilesafe;
import android.app.Activity;
import android.content.Intent;
import android.content.pm.PackageInfo;
```

```java
import android.content.pm.PackageManager.NameNotFoundException;
import android.os.Bundle;
import android.text.TextUtils;
import android.view.KeyEvent;
import android.view.View;
import android.widget.EditText;
import android.widget.ImageView;
import android.widget.TextView;
import android.widget.Toast;
public class EnterPwdActivity extends Activity {
    //密码输入框
    private EditText et_password;
    //应用名称
    private TextView tv_name;
    //应用图标
    private ImageView iv_icon;
    //应用包名
    private String packname;
    @Override
    protected void onCreate(Bundle savedInstanceState) {
        super.onCreate(savedInstanceState);
        setContentView(R.layout.enter_pwd);
        et_password = (EditText) findViewById(R.id.et_password);
        //获取到激活当前Activity的意图（WatchDogService1中的pwdintent）
        Intent intent = getIntent();
        //获取到意图中存入的数据（要进入被锁定的应用的包名）
        packname = intent.getStringExtra("packname");
        tv_name = (TextView) findViewById(R.id.tv_enterpwd_name);
        iv_icon = (ImageView) findViewById(R.id.iv_enterpwd_icon);
        try {
            //根据包名获取到包信息对象
   PackageInfo info = getPackageManager().getPackageInfo(packname, 0);
            //info.applicationInfo.loadLabel(getPackageManager())获取到该包
            名的应用程序所对应的应用名称 tv_name.setText(info.applicationInfo.
            loadLabel(getPackageManager()));
            //info.applicationInfo.loadIcon(getPackageManager())获取到该
            包名的应用程序所对应的应用图标
 iv_icon.setImageDrawable(info.applicationInfo.loadIcon(getPackageMa
                        nager()));
        } catch (NameNotFoundException e) {
            e.printStackTrace();
        }
    }
    /**
     * 单击"确定"按钮时执行的方法
     */
```

```java
public void enterPassword(View view){
    //获取到输入框中的密码,并将密码前后的空格清除掉
    String pwd = et_password.getText().toString().trim();
    //判断输入的密码是否为空
    if(TextUtils.isEmpty(pwd)){
        Toast.makeText(this, "密码不能为空", 0).show();
        return ;
    }
    //判断密码是否为123(正确密码,没有提供设置密码的界面,这里简单处理一下)
    if("123".equals(pwd)){
        finish();

    }else{
        Toast.makeText(this, "密码不正确", 0).show();
        return ;
    }
}
/**
 * 当进入当前的界面后,屏蔽掉 Back 键
 */
@Override
public boolean onKeyDown(int keyCode, KeyEvent event) {
    if(event.getAction()==KeyEvent.ACTION_DOWN&&event.getKeyCode()==KeyEvent.KEYCODE_BACK){
        return true;//消费掉当前的 Back 键
    }
    return super.onKeyDown(keyCode, event);
}
```

代码解析:

onKeyDown (int keyCode, KeyEvent event)

该方法是在按下手机上的任意一个按键时执行的,这里用于当用户进入输入密码界面时,不允许用户按 Back 键后退。

EnterPwdActivity 的实现逻辑:获取到 WatchDogService1 中启动 EnterPwdActivity 的意图对象→获取到该意图中的数据(要进入被锁定的应用的包名)→通过包名来获取被锁定的应用程序的信息(应用名称、图标)→将这些信息显示在当前界面对应的位置上。在输入密码时,我们没有设置设置密码的界面,在这里只是简单地处理一下。当密码输入正确后,会"finish()"掉当前的界面。

测试运行:在清单文件中为各个组件配置好组件信息及需要的权限信息→锁定"计算器"应用程序→在设置中心中重新开启程序锁服务→开启服务后,按 Back 键退出手机卫士(读者在测试时需要按照该顺序操作,后面会一步步地修复其中的 bug)→当打开计算器应用时,界面如图 3-28 所示。

图 3-28

当正确输入密码后,当前的 Activity 会执行 finish()方法,当前的 Activity 将会被移出栈顶,此时进入计算器界面,但是,计算器界面也只是一闪而过,此时再次进入了输入密码的界面。出现这种现象的原因如下:

当要打开被锁定的计算器应用时,该应用将处于任务栈栈顶(可以看到主界面)→此时看门狗也就获取到了计算器应用对应的包名,并发现该应用的包名是一个被锁定的应用→立即进入输入密码界面,输入正确密码后,执行 finish()方法→输入密码的 Activity 从任务栈顶上移除掉,计算机对应的 Activity 又会位于栈顶,此时,看门狗发现此时的任务栈栈顶应用的包名又是被锁定的,所以会再次进入输入密码的界面。

解决方案:当正确输入密码后,应当通知看门狗,临时停止对该应用程序的保护。此时,WatchDogService1.java 中的业务代码如下:

```java
package com.guoshisp.mobilesafe.service;
import java.util.ArrayList;
import java.util.List;
import android.app.ActivityManager;
import android.app.ActivityManager.RunningTaskInfo;
import android.app.Service;
import android.content.Intent;
import android.os.Binder;
import android.os.IBinder;
import android.util.Log;
import com.guoshisp.mobilesafe.EnterPwdActivity;
import com.guoshisp.mobilesafe.IService;
import com.guoshisp.mobilesafe.db.dao.AppLockDao;
public class WatchDogService1 extends Service {
    protected static final String TAG = "WatchDogService";
```

```java
//是否要停止掉看门狗服务。true 表示继续运行，false 表示停止运行
boolean flag;
//要进入一个已被锁定的应用程序前，需要输入正确的密码后才可以进入。这是一个用于激
  活输入密码的界面
private Intent pwdintent;
//将所有已被锁定的应用程序的包名存放在该集合缓存中
private List<String> lockPacknames;
//操作数据库的对象
private AppLockDao dao;
//存放临时需要被保护的应用程序包名
private List<String> tempStopProtectPacknames;
//要返回的给 EnterPwdActivity 中的 ServiceConnection 对象
private MyBinder binder;
//返回到 EnterPwdActivity 中的 ServiceConnection 对象中 onServiceConnected
  (ComponentName name, IBinder service)方法的第二个参数
@Override
public IBinder onBind(Intent intent) {
    binder = new MyBinder();
    return binder;
}
private class MyBinder extends Binder implements IService{
    public void callTempStopProtect(String packname) {
        tempStopProtect(packname);
    }
}
//临时停止保护一个被锁定的应用程序的方法
public void tempStopProtect(String packname){
    //将需要临时停止保护的程序的包名添加到对应的集合中
    tempStopProtectPacknames.add(packname);
}
@Override
public void onCreate() {
    super.onCreate();
    dao = new AppLockDao(this);
    //将看门狗服务的标记设置为 true，让其一直在后台运行
    flag = true;
    tempStopProtectPacknames = new ArrayList<String>();
    //从程序锁对应的数据库中取出所有应用程序的包名
    lockPacknames = dao.findAll();
    pwdintent = new Intent(this,EnterPwdActivity.class);
    //因为服务本身没有任务栈，如果要开启一个需要在任务栈中运行的 Activity，需要
      为该 Activity 创建一个任务栈
    pwdintent.setFlags(Intent.FLAG_ACTIVITY_NEW_TASK);
    //开启一个线程不断地运行看门狗服务
    new Thread() {
```

```java
        public void run() {
            //设置一个死循环,如果为true,则一直运行
            while (flag){
                //获取一个Activity的管理器,ActivityManager可以动态地观察到当前存在哪些进程
                ActivityManager am = (ActivityManager) getSystemService
                            (ACTIVITY_SERVICE);
                //获取到当前正在栈顶运行的Activity
                RunningTaskInfo taskinfo = am.getRunningTasks(1).get(0);
                //获取到当前任务栈顶程序所对应的包名
                String packname = taskinfo.topActivity.getPackageName();
                Log.i(TAG,packname);
                //判断当前栈顶应用程序对应的包名是否是临时被保护的程序
                if(tempStopProtectPacknames.contains(packname)){
                    try {
                        //看门狗服务非常耗电,这里用于让该服务暂停200毫秒
                        Thread.sleep(200);
                    } catch (InterruptedException e) {
                        e.printStackTrace();
                    }
        //当前栈顶应用程序对应的包名是临时被保护的程序,则跳出当前的if语句,继续执行while循环
                    continue;
                }
                //将任务栈顶的程序的包名信息存入意图中(以键值对的形式存入,可
                    以在被激活的Activity中通过getIntent()来获取该意图,然后
                    再获取意图对象中的数据)
                pwdintent.putExtra("packname", packname);
                //判断运行在栈顶的程序所对应的包名是否是已锁定的应用程序
                if(lockPacknames.contains(packname)){
                //发现当前应用程序为已锁定的应用程序,需要进入输入密码的界面
                    startActivity(pwdintent);
                }
                try {
                    //看门狗服务非常耗电,这里用于让该服务暂停200毫秒
                    Thread.sleep(200);
                } catch (InterruptedException e) {
                    e.printStackTrace();
                }
            }
        };
    }.start();
}
//当服务被停止时,应停止看门狗
@Override
public void onDestroy() {
    flag = false;
```

```java
        super.onDestroy();
    }
}
```

EnterPwdActivity.java 中的业务代码如下：

```java
package com.guoshisp.mobilesafe;
import android.app.Activity;
import android.content.ComponentName;
import android.content.Intent;
import android.content.ServiceConnection;
import android.content.pm.PackageInfo;
import android.content.pm.PackageManager.NameNotFoundException;
import android.os.Bundle;
import android.os.IBinder;
import android.text.TextUtils;
import android.view.KeyEvent;
import android.view.View;
import android.widget.EditText;
import android.widget.ImageView;
import android.widget.TextView;
import android.widget.Toast;
import com.guoshisp.mobilesafe.service.WatchDogService1;
public class EnterPwdActivity extends Activity {
    //密码输入框
    private EditText et_password;
    //应用名称
    private TextView tv_name;
    //应用图标
    private ImageView iv_icon;
    //用于启动看门狗服务的意图对象
    private Intent serviceIntent;
    //停止保护一个应用程序（接口）
    private IService iService;
    //连接服务时的一个对象（在绑定服务时需要传入）
    private MyConn conn;
    //应用包名
    private String packname;
    @Override
    protected void onCreate(Bundle savedInstanceState) {
        super.onCreate(savedInstanceState);
        setContentView(R.layout.enter_pwd);
        et_password = (EditText) findViewById(R.id.et_password);
        //获取到激活当前 Activity 的意图（WatchDogService1 中的 pwdintent）
        Intent intent = getIntent();
        //获取到意图中存入的数据（要进入被锁定的应用的包名）
        packname = intent.getStringExtra("packname");
```

```java
        tv_name = (TextView) findViewById(R.id.tv_enterpwd_name);
        iv_icon = (ImageView) findViewById(R.id.iv_enterpwd_icon);
        serviceIntent = new Intent(this,WatchDogService1.class);
        conn = new MyConn();
        //绑定服务（非 startService()）。执行服务中的 onCreate-->onBind 方法（该
            方法的返回值不能为 null）
        bindService(serviceIntent, conn, BIND_AUTO_CREATE);
        try {
            //根据包名获取到包信息对象
    PackageInfo info = getPackageManager().getPackageInfo(packname, 0);
            //info.applicationInfo.loadLabel(getPackageManager())获取到该
                包名的应用程序所对应的应用名称
tv_name.setText(info.applicationInfo.loadLabel(getPackageManager()));
            //info.applicationInfo.loadIcon(getPackageManager())获取到该
                包名的应用程序所对应的应用图标
        iv_icon.setImageDrawable(info.applicationInfo.loadIcon
                            (getPackageManager()));
        } catch (NameNotFoundException e) {
            e.printStackTrace();
        }
    }
    private class MyConn implements ServiceConnection{
        //在操作者连接一个服务成功时被调用
        public void onServiceConnected(ComponentName name, IBinder service) {
            //因为返回的 IBinder 实现了 iService 接口（向上转型）
            iService = (IService) service;
        }
        //在服务崩溃或被杀死导致的连接中断时被调用，而如果我们自己解除绑定时则不会被调用
        public void onServiceDisconnected(ComponentName name) {

        }
    }
    @Override
    protected void onDestroy() {
        super.onDestroy();
        //解除绑定
        unbindService(conn);
    }
    /**
     * 单击"确定"按钮时执行的方法
     */
    public void enterPassword(View view){
        //获取到输入框中的密码，并将密码前后的空格清除
        String pwd = et_password.getText().toString().trim();
        //判断输入的密码是否为空
        if(TextUtils.isEmpty(pwd)){
```

```
            Toast.makeText(this, "密码不能为空", 0).show();
            return ;
        }
        //判断密码是否为123（正确密码，没有提供设置密码的界面，这里简单处理一下）
        if("123".equals(pwd)){
            //通知看门狗，临时停止对packname的保护
            iService.callTempStopProtect(packname);
            finish();
        }else{
            Toast.makeText(this, "密码不正确", 0).show();
            return ;
        }
    }
    /**
     * 当进入当前的界面后，屏蔽掉Back键
     */
    @Override
    public boolean onKeyDown(int keyCode, KeyEvent event) {
        if(event.getAction()==KeyEvent.ACTION_DOWN&&event.getKeyCode()==Key
                    Event.KEYCODE_BACK){
            return true;//消费掉当前的Back键
        }
        return super.onKeyDown(keyCode, event);
    }
}
```

使看门狗服务停止对packname保护的接口IService位于包"com.guoshisp.mobilesafe"下，IService.java如下：

```
package com.guoshisp.mobilesafe;
public interface IService {
    public void callTempStopProtect(String packname);
}
```

思路解析：当"设置中心"将看门狗服务启动起来后，服务已经完成了初始化工作（此时，tempStopProtectPacknames集合中的没有添加任何元素），当用于再次进入"计算器"应用时，进入输入密码的界面EnterPwdActivity中，在EnterPwdActivity完成初始化的过程中，进行了绑定看门狗服务的操作（bindService(serviceIntent,conn, BIND_AUTO_CREATE)）。绑定服务时，会执行看门狗服务中的onBind (Intent intent)方法，如果绑定成功，onBind (Intent intent)方法会返回一个IBinder给EnterPwdActivity中的ServiceConnection对象中onServiceConnected(ComponentName name, IBinder service)方法的第二个参数，而返回的这个IBinder对象已经实现了接口IService，所以我们得到返回的IBinder对象后可以直接调用接口中的方法：iService.callTempStopProtect(packname)。而在callTempStopProtect(packname)方法中执行的是 tempStopProtect(packname)方法，在 tempStopProtect(packname)方法中执行的是tempStopProtectPacknames.add(packname)，此时，看门狗服务中的 tempStopProtectPacknames

集合中就多了一个元素，该元素正是需要被暂停保护的应用程序。当看门狗服务再次执行 if(tempStopProtectPacknames.contains(packname))时，会执行 continue 语句，跳出当前的 if 语句，重新执行 while 循环，也就是说，只要执行了 continue 语句，当前 if 语句下面的都不会被执行，也就不会执行 startActivity(pwdintent)，而是直接进入"计算器"应用程序中。

3.7.2 程序锁中的 bug 解决方案

下面，来解决程序锁中的几个 bug：

（1）当我们在"设置中心"中启动程序锁服务后，然后通过"高级工具"进入程序锁界面，此时，即使再次锁定某个应用程序，或者解锁应用程序，应用程序都不会被成功地锁定或者解锁。原因在于：虽然数据库中的信息可以改变（增加、删除），但是在看门狗服务中我们只对数据库进行了一次查询的操作，当数据库中的数据在此次查询之后发生改变，那么数据也就不能够保持同步了。

解决方案：使用自定义的内容提供者来观察数据库中的数据（使用内容观察者将会更加方便，这里，主要是方便读者巩固一下四大组件之一——内容提供者）。在包"com.guoshisp.mobilesafe.provider"下创建一个 AppLockDBProvider，AppLockDBProvider.java 代码如下：

```java
package com.guoshisp.mobilesafe.provider;
import android.content.ContentProvider;
import android.content.ContentValues;
import android.content.UriMatcher;
import android.database.Cursor;
import android.net.Uri;
import com.guoshisp.mobilesafe.db.dao.AppLockDao;
/**
 * 在内容提供者中，只需要关心对数据库的增、删操作
 * @author Administrator
 *
 */
public class AppLockDBProvider extends ContentProvider {
    private static final int ADD = 1;
    //content://com.guoshisp.applock/ADD
    //content://com.guoshisp.applock/DELETE
    //获取操作数据库的对象
    private AppLockDao dao;
    //定义匹配码
    private static final int DELETE = 2;
    public static UriMatcher matcher = new UriMatcher(UriMatcher.NO_MATCH);
    //定义匹配路径
    static {
        //匹配 Uri。参数一：主机名；参数二：指定数据库中的表名或者一些业务逻辑（add、delete 等）
        //参数三：匹配码。即执行匹配时判断匹配是否正确：matcher.match(uri)
```

```java
        matcher.addURI("com.guoshisp.applock", "ADD", ADD);
        matcher.addURI("com.guoshisp.applock", "DELETE", DELETE);
    }
    @Override
    public boolean onCreate() {
        dao = new AppLockDao(getContext());
        return false;
    }
    @Override
    public Cursor query(Uri uri, String[] projection, String selection,
            String[] selectionArgs, String sortOrder) {
        return null;
    }
    @Override
    public String getType(Uri uri) {
        //TODO Auto-generated method stub
        return null;
    }
    @Override
    public Uri insert(Uri uri, ContentValues values) {
        //匹配URI
        int result = matcher.match(uri);
        //判断是否是添加的匹配操作
        if (result == ADD) {
            //获取到添加的包名（ContentValues是在Item被单击是添加的）
            String packname = values.getAsString("packname");
            //添加到数据库中
            dao.add(packname);
            //发布内容的变化通知
            getContext().getContentResolver().notifyChange(uri, null);
        }
        return null;
    }
    @Override
    public int delete(Uri uri, String selection, String[] selectionArgs) {
        int result = matcher.match(uri);
        if (result == DELETE) {
            dao.delete(selectionArgs[0]);
            //发布内容的变化通知
            getContext().getContentResolver().notifyChange(uri, null);
        }
        return 0;
    }
    @Override
    public int update(Uri uri, ContentValues values, String selection,
```

```
            String[] selectionArgs) {
        return 0;
    }
}
```

定义好该组件之后，需要在清单文件中为该组件配置一下组件信息：

```
<provider
        android:name=".provider.AppLockDBProvider"
        android:authorities="com.guoshisp.applock" >
</provider>
```

我们必须要带上属性 android:authorities，后面的内容可以随便添加。android: authorities 是用于指定内容提供者的主机名，以后要访问数据库，Uri 的匹配格式应该为：content://com.guoshisp.applock/xxx。

完成内容提供者的设置后，在单击 Item 时，不能够通过 AppInfoProvider 来操作数据库，应该使用自己定义好的内容提供者来操作数据库。此时，**AppLockActivity.java** 中的最终业务代码如下：

```java
package com.guoshisp.mobilesafe;
import java.util.List;
import android.app.Activity;
import android.content.ContentValues;
import android.net.Uri;
import android.os.Bundle;
import android.os.Handler;
import android.view.View;
import android.view.ViewGroup;
import android.view.animation.Animation;
import android.view.animation.TranslateAnimation;
import android.widget.AdapterView;
import android.widget.AdapterView.OnItemClickListener;
import android.widget.BaseAdapter;
import android.widget.ImageView;
import android.widget.LinearLayout;
import android.widget.ListView;
import android.widget.TextView;
import com.guoshisp.mobilesafe.db.dao.AppLockDao;
import com.guoshisp.mobilesafe.domain.AppInfo;
import com.guoshisp.mobilesafe.engine.AppInfoProvider;
public class AppLockActivity extends Activity {
//展示手机中的所有应用
    private ListView lv_applock;
//ProgressBar 和 TextView 对应的父控件，用于控制 ProgressBar 和 TextView 的显示
    private LinearLayout ll_loading;
    //获取手机中已安装的应用程序
    private AppInfoProvider provider;
```

```java
//存放当前手机上所有应用程序的信息
private List<AppInfo> appinfos;
//操作存放已锁定的应用程序的数据库
private AppLockDao dao;
//存放所有已经被锁定的应用程序的包名信息
private List<String> lockedPacknames;
//处理子线程中获取到的当前手机中所有应用程序
private Handler handler = new Handler(){
    public void handleMessage(android.os.Message msg) {
        ll_loading.setVisibility(View.INVISIBLE);
        //为ListView适配数据
        lv_applock.setAdapter(new AppLockAdapter());
    };
};
@Override
protected void onCreate(Bundle savedInstanceState) {
    setContentView(R.layout.app_lock);
    super.onCreate(savedInstanceState);
    provider = new AppInfoProvider(this);
    lv_applock = (ListView) findViewById(R.id.lv_applock);
    ll_loading = (LinearLayout) findViewById(R.id.ll_applock_loading);
    dao =new AppLockDao(this);
    //从数据库中获取到所有被锁定的应用程序包名
    lockedPacknames = dao.findAll();
    //正在从数据库中获取数据时,应该显示ProgressBar和TextView对应的"正在加
      载…"字样
    ll_loading.setVisibility(View.VISIBLE);
    //开启一个子线程获取手机中所有应用程序的信息
    new Thread(){
        public void run() {
            appinfos = provider.getInstalledApps();
            //向主线程中发送一个空消息,通知主线程更新数据
            handler.sendEmptyMessage(0);
        };
    }.start();
    //为ListView中的Item设置单击事件的监听器
    lv_applock.setOnItemClickListener(new OnItemClickListener() {
        public void onItemClick(AdapterView<?> parent, View view,
                int position, long id) {
            //获取当前Item的对象
            AppInfo appinfo = (AppInfo) lv_applock.getItemAtPosition
                        (position);
            //获取到当前Item对象的包名信息
            String packname = appinfo.getPackname();
            //查找到Item对应的锁控件(ImageView)
            ImageView iv = (ImageView) view.findViewById(R.id.iv_
```

```java
                            applock_status);
                //设置一个左右移动的动画
                TranslateAnimation ta = new TranslateAnimation(Animation.
                            RELATIVE_TO_SELF, 0, Animation.
                            RELATIVE_TO_SELF, 0.2f, Animation.
                            RELATIVE_TO_SELF, 0, Animation.
                            RELATIVE_TO_SELF, 0);
                //设置动画播放的时长(毫秒)
                ta.setDuration(200);
                //判断当前的Item是否处于锁定状态,如果是,则应该解锁,否则应该加锁
                if(lockedPacknames.contains(packname)){//锁定状态
                    //dao.delete(packname);
                    //采用内容提供者来观察数据库中的数据变化
                    Uri uri = Uri.parse("content://com.guoshisp.applock
                            DELETE");
                    getContentResolver().delete(uri, null, new String[]
                                {packname});
                    //解锁
                    iv.setImageResource(R.drawable.unlock);
//将当前应用程序的包名从集合(存放已锁定应用程序的包名)中移除,以便界面的刷新
                    lockedPacknames.remove(packname);
                }else{//未锁定状态
                    //dao.add(packname);
                    Uri uri = Uri.parse("content://com.guoshisp.applock/ADD");
                    ContentValues values = new ContentValues();
                    values.put("packname", packname);
                    getContentResolver().insert(uri, values);
                    iv.setImageResource(R.drawable.lock);
                    lockedPacknames.add(packname);
                }
                //为当前的Item播放动画
                view.startAnimation(ta);
            }
        });
    }
    //自定义适配器对象
    private class AppLockAdapter extends BaseAdapter{
        public int getCount() {
            return appinfos.size();
        }
        public Object getItem(int position) {
            return appinfos.get(position);
        }
        public long getItemId(int position) {
            return position;
        }
```

```java
        public View getView(int position, View convertView, ViewGroup
                    parent) {
            View view;
            ViewHolder holder;
            //复用历史缓存的View对象
            if(convertView==null){
                view = View.inflate(getApplicationContext(),R.layout.app_
                        lock_item, null);
                holder = new ViewHolder();
                holder.iv_icon = (ImageView)view.findViewById(R.id.iv_
                        applock_icon);
                holder.iv_status = (ImageView)view.findViewById(R.id.iv_
                        applock_status);
                holder.tv_name = (TextView)view.findViewById(R.id.tv_
                        applock_appname);
                view.setTag(holder);
            }else{//为View做一个标记，以便复用
                view = convertView;
                holder = (ViewHolder) view.getTag();
            }
            //获取到当前应用程序对象
            AppInfo appInfo = appinfos.get(position);
            holder.iv_icon.setImageDrawable(appInfo.getAppicon());
            holder.tv_name.setText(appInfo.getAppname());
        //查看被当前的Item是否是被绑定的应用，以此来为Item设置对应的锁（锁定或未锁定）
            if(lockedPacknames.contains(appInfo.getPackname())){
                holder.iv_status.setImageResource(R.drawable.lock);
            }else{
                holder.iv_status.setImageResource(R.drawable.unlock);
            }
            return view;
        }
    }
    //View对应的View对象只会在堆内存中存在一份，所有的Item都公用该View
    public static class ViewHolder{
        ImageView iv_icon;
        ImageView iv_status;
        TextView tv_name;
    }
}
```

在看门狗服务中，我们需要了解到数据库的变化情况，此时需要在该服务中注册一个内容观察者来观察数据库中的数据的改变。此时，WatchDogService1.java中的业务代码如下：

```java
package com.guoshisp.mobilesafe.service;
import java.util.ArrayList;
import java.util.List;
```

```java
import android.app.ActivityManager;
import android.app.ActivityManager.RunningTaskInfo;
import android.app.Service;
import android.content.Intent;
import android.database.ContentObserver;
import android.net.Uri;
import android.os.Binder;
import android.os.Handler;
import android.os.IBinder;
import android.util.Log;
import com.guoshisp.mobilesafe.EnterPwdActivity;
import com.guoshisp.mobilesafe.IService;
import com.guoshisp.mobilesafe.db.dao.AppLockDao;
public class WatchDogService1 extends Service {
    protected static final String TAG = "WatchDogService";
    //是否要停止掉看门狗服务。true 表示继续运行，false 表示停止运行
    boolean flag;
    //要进入一个已被锁定的应用程序前，需要输入正确的密码后才可以进入。这是一个用于激
      活输入密码的界面
    private Intent pwdintent;
    //将所有已被锁定的应用程序的包名存放在该集合缓存中
    private List<String> lockPacknames;
    //操作数据库的对象
    private AppLockDao dao;
    //存放临时需要被保护的应用程序包名
    private List<String> tempStopProtectPacknames;
    //返回到 EnterPwdActivity 中的 ServiceConnection 对象中 onServiceConnected
      (ComponentName name, IBinder service)方法的第二个参数
    private MyBinder binder;
    //内容观察者
    private MyObserver observer;
    @Override
    public IBinder onBind(Intent intent) {
        binder = new MyBinder();
        return binder;
    }
    private class MyBinder extends Binder implements IService{
        public void callTempStopProtect(String packname) {
            tempStopProtect(packname);
        }
    }
    //临时停止保护一个被锁定的应用程序的方法
    public void tempStopProtect(String packname){
        //将需要临时停止保护的程序的包名添加到对应的集合中
        tempStopProtectPacknames.add(packname);
    }
```

```java
@Override
public void onCreate() {
    //设置要匹配的Uri路径
    Uri uri = Uri.parse("content://com.guoshisp.applock/");
    observer = new MyObserver(new Handler());
    //第二个参数如果为true, Uri中的content://com.guoshisp.applock/匹配
      正确即可感应到, 后面的 (ADD或DELETE) 不用继续在匹配下去
    getContentResolver().registerContentObserver(uri, true, observer);
    super.onCreate();
    dao = new AppLockDao(this);
    //将看门狗服务的标记设置为true, 让其一直在后台运行
    flag = true;
    tempStopProtectPacknames = new ArrayList<String>();
    //从程序锁对应的数据库中取出所有应用程序的包名
    lockPacknames = dao.findAll();
    pwdintent = new Intent(this,EnterPwdActivity.class);
    //因为服务本身没有任务栈, 如果要开启一个需要在任务栈中运行的Activity,
      需要为该Activity创建一个任务栈
    pwdintent.setFlags(Intent.FLAG_ACTIVITY_NEW_TASK);
    //开启一个线程不断的运行看门狗服务
    new Thread() {
        public void run() {
            //设置一个死循环, 如果为true, 则一直运行
            while (flag){
//获取一个Activity的管理器, ActivityManager可以动态地观察当前存在哪些进程
                ActivityManager am = (ActivityManager) getSystemService
                                (ACTIVITY_SERVICE);
                //获取到当前正在栈顶运行的Activity
                RunningTaskInfo taskinfo = am.getRunningTasks(1).get(0);
                //获取到当前任务栈顶程序所对应的包名
                String packname = taskinfo.topActivity.getPackageName();
                Log.i(TAG,packname);
                //判断当前栈顶应用程序对应的包名是否是临时被保护的程序
                if(tempStopProtectPacknames.contains(packname)){
                    try {
                        //看门狗服务非常耗电, 这里用于让该服务暂停200毫秒
                        Thread.sleep(200);
                    } catch (InterruptedException e) {
                        e.printStackTrace();
                    }
                    //当前栈顶应用程序对应的包名是临时被保护的程序, 则跳出当前
                      的if语句, 继续执行while循环
                    continue;
                }
                //将任务栈顶的程序的包名信息存入意图中 (以键值对的形式存入, 可
                  以在被激活的Activity中通过getIntent()来获取该意图, 然后
```

再获取意图对象中的数据）
```
                pwdintent.putExtra("packname", packname);
                //判断运行在栈顶的程序所对应的包名是否是已锁定的应用程序
                if(lockPacknames.contains(packname)){
                //发现当前应用程序为已锁定的应用程序，需要进入输入密码的界面
                    startActivity(pwdintent);
                }
                try {
                    //看门狗服务非常耗电，这里用于让该服务暂停200毫秒
                    Thread.sleep(200);
                } catch (InterruptedException e) {
                    e.printStackTrace();
                }
            }
        };
    }.start();
}
//当服务被停止时，应停止看门狗，同时将内容观察者反注册掉
@Override
public void onDestroy() {
    flag = false;
    //将内容观察者反注册掉
    getContentResolver().unregisterContentObserver(observer);
    observer = null;
    super.onDestroy();
}
private class MyObserver extends ContentObserver{
    public MyObserver(Handler handler) {
        super(handler);
    }
    //当对应的Uri中的数据发生改变时调用该方法
    @Override
    public void onChange(boolean selfChange) {
        //重新从数据库中获取数据
        lockPacknames = dao.findAll();
        super.onChange(selfChange);
    }
}
}
```

测试运行：首先开启"设置中心"中的程序锁服务，然后到"高级工具"的程序锁中将浏览器应用锁上，接着按Back键退出手机安全卫士，打开浏览器，界面效果如图3-29所示。

到"设置中心"中将程序锁服务开启，然后将"相机"应用锁定，此时不按Back键返回，直接按Home键退出到桌面，然后打开相机，当正确地输入密码后，界面如图3-30所示。

这个界面显然不是用户希望出现的，我们希望用户在正确地输入密码后，应该直接进入到

对应的应用中。出现该 bug 的原因在于 Activity 的任务栈的问题：一般来说，每一个应用程序都会有自己的一个任务栈（手机安全卫士有自己的任务栈）。当启动输入密码界面的 Activity 时，系统会检查手机安全卫士的应用是否已经有了任务栈，如果有，就直接将输入密码的 Activity 加入到手机安全卫士的任务栈栈顶。由于当时是在程序锁界面直接按 Home 键退到桌面的，所以，程序锁所对应的任务栈此时位于任务栈的栈顶。当我们进入输入密码的界面时，此时的输入密码界面对应的 Activity 位于该任务栈的栈顶（此时，程序锁对应的 Activity 在任务栈的位置仅次于输入密码对应的 Activity），所以，当正确输入密码后，输入密码的 Activity 代码中执行了 finish()方法，将该 Activity 移出栈顶。此时，程序锁界面对应的 Activity 再次回到了任务栈栈顶，一般位于任务栈栈顶的 Activity 就是当前正在显示的界面。

图 3-29　　　　　　　　　　　图 3-30

解决方案：将输入密码的 Activity 放在单独的任务栈中。只需要配置一下该 Activity 的启动模式：

```xml
<activity
    android:name=".EnterPwdActivity"
    android:launchMode="singleInstance" />
```

将"联系人"或其他的应用锁定起来，当单击"联系人"应用时会进入输入密码的界面，此时不输入密码，直接按 Home 键退出，然后长按 Home 键打开最近使用的那个应用（其实就是"手机安全卫士"应用），进入时，显示的是进入联系人应用时需要输入密码解锁的界面。这一点对于开发人员来说是可以理解的，但对于用户来说，这是一个比较奇怪的现象。

解决方案：不让需要输入密码的 Activity 在最近任务栏菜单中显示。只需要配置一下该 Activity 的启动模式：

```xml
<activity
    android:name=".EnterPwdActivity"
    android:excludeFromRecents="true"
```

```
                android:launchMode="singleInstance" />
```

应用程序在解锁后,就成为被暂停保护的应用程序了,以后再次进入时不需要输入密码即可进入。如果是其他人使用我们的手机,那么他同样也可以进入该应用程序,这不是我们希望的。除非我们重新开启程序锁服务,这显然是很不方便的。

解决方案:可以在屏幕处于锁屏状态时继续对已锁定的应用程序保护(清空暂停被保护程序的集合)。锁屏操作可以通过一个广播来实现,我们将其定义在 WatchDogService1,此时,最终的 WatchDogService1.java 业务代码如下:

```java
package com.guoshisp.mobilesafe.service;
import java.util.ArrayList;
import java.util.List;
import android.app.ActivityManager;
import android.app.ActivityManager.RunningTaskInfo;
import android.app.Service;
import android.content.BroadcastReceiver;
import android.content.Context;
import android.content.Intent;
import android.content.IntentFilter;
import android.database.ContentObserver;
import android.net.Uri;
import android.os.Binder;
import android.os.Handler;
import android.os.IBinder;
import android.util.Log;
import com.guoshisp.mobilesafe.EnterPwdActivity;
import com.guoshisp.mobilesafe.IService;
import com.guoshisp.mobilesafe.db.dao.AppLockDao;
public class WatchDogService1 extends Service {
    protected static final String TAG = "WatchDogService";
    //是否要停止掉看门狗服务。true 表示继续运行,false 表示停止运行
    boolean flag;
    //要进入一个已被锁定的应用程序前,需要输入正确的密码后才可以进入。这是一个用于激
      活输入密码的界面
    private Intent pwdintent;
    //将所有已被锁定的应用程序的包名存放在该集合缓存中
    private List<String> lockPacknames;
    //操作数据库的对象
    private AppLockDao dao;
    //存放临时需要被保护的应用程序包名
    private List<String> tempStopProtectPacknames;
    //返回到 EnterPwdActivity 中的 ServiceConnection 对象中 onServiceConnected
      (ComponentName name, IBinder service)方法的第二个参数
    private MyBinder binder;
    //内容观察者
    private MyObserver observer;
```

```java
//锁屏的广播接收者
private LockScreenReceiver receiver;
@Override
public IBinder onBind(Intent intent) {
    binder = new MyBinder();
    return binder;
}
private class MyBinder extends Binder implements IService{
    public void callTempStopProtect(String packname) {
        tempStopProtect(packname);
    }
}
//临时停止保护一个被锁定的应用程序的方法
public void tempStopProtect(String packname){
    //将需要临时停止保护的程序的包名添加到对应的集合中
    tempStopProtectPacknames.add(packname);
}
@Override
public void onCreate() {
    //设置要匹配的Uri路径
    Uri uri = Uri.parse("content://com.guoshisp.applock/");
    observer = new MyObserver(new Handler());
    //第二个参数如果为true, Uri中的content://com.guoshisp.applock/匹配
    //  正确即可感应到,后面的(ADD或DELETE)不用继续在匹配下去
    getContentResolver().registerContentObserver(uri, true, observer);
    //以代码动态注册一个广播接收者
    IntentFilter filter = new IntentFilter();
    //优先级
    filter.setPriority(1000);
    //要接收的动作
    filter.addAction(Intent.ACTION_SCREEN_OFF);
    receiver = new LockScreenReceiver();
    //采用代码动态的注册广播接受者
    registerReceiver(receiver, filter);
    super.onCreate();
    dao = new AppLockDao(this);
    //将看门狗服务的标记设置为true,让其一直在后台运行
    flag = true;
    tempStopProtectPacknames = new ArrayList<String>();
    //从程序锁对应的数据库中取出所有应用程序的包名
    lockPacknames = dao.findAll();
    pwdintent = new Intent(this,EnterPwdActivity.class);
    //因为服务本身没有任务栈,如果要开启一个需要在任务栈中运行的Activity,
    //  需要为该Activity创建一个任务栈
    pwdintent.setFlags(Intent.FLAG_ACTIVITY_NEW_TASK);
    //开启一个线程不断地运行看门狗服务
    new Thread() {
```

```java
        public void run() {
            //设置一个死循环，如果为true，则一直运行
            while (flag){
//获取一个Activity的管理器，ActivityManager可以动态地观察到当前存在哪些进程
                ActivityManager am = (ActivityManager) getSystemService
                            (ACTIVITY_SERVICE);
                //获取到当前正在栈顶运行的Activity
                RunningTaskInfo taskinfo = am.getRunningTasks(1).get(0);
                //获取到当前任务栈顶程序所对应的包名
                String packname = taskinfo.topActivity.getPackageName();
                Log.i(TAG,packname);
                //判断当前栈顶应用程序对应的包名是否是临时被保护的程序
                if(tempStopProtectPacknames.contains(packname)){
                    try {
                        //看门狗服务非常耗电，这里用于让该服务暂停200毫秒
                        Thread.sleep(200);
                    } catch (InterruptedException e) {
                        e.printStackTrace();
                    }
                    //当前栈顶应用程序对应的包名是临时被保护的程序，则跳出当前
                        的if语句，继续执行while循环
                    continue;
                }
                //将任务栈顶的程序的包名信息存入意图中（以键值对的形式存入，可
                    以在被激活的Activity中通过getIntent()来获取该意图，然后
                    再获取意图对象中的数据）
                pwdintent.putExtra("packname", packname);
                //判断运行在栈顶的程序所对应的包名是否是已锁定的应用程序
                if(lockPacknames.contains(packname)){
                    //发现当前应用程序为已锁定的应用程序，需要进入输入密码的界面
                    startActivity(pwdintent);
                }
                try {
                    //看门狗服务非常耗电，这里用于让该服务暂停200毫秒
                    Thread.sleep(200);
                } catch (InterruptedException e) {
                    e.printStackTrace();
                }
            }
        };
    }.start();
}
//当服务被停止时，我们应停止看门狗继续运行，同时将内容观察者反注册掉，反注册掉
    广播接收者
@Override
```

```java
    public void onDestroy() {
        flag = false;
        //将内容观察者反注册掉
        getContentResolver().unregisterContentObserver(observer);
        observer = null;
        //反注册掉广播接收者
        unregisterReceiver(receiver);
        super.onDestroy();
    }
    private class MyObserver extends ContentObserver{
        public MyObserver(Handler handler) {
            super(handler);
        }
        //当对应的Uri中的数据发生改变时调用该方法
        @Override
        public void onChange(boolean selfChange) {
            //重新从数据库中获取数据
            lockPacknames = dao.findAll();
            super.onChange(selfChange);
        }
    }
    private class LockScreenReceiver extends BroadcastReceiver{
        @Override
        public void onReceive(Context context, Intent intent) {
            Log.i(TAG,"锁屏了");
            //清空集合，继续保护
            tempStopProtectPacknames.clear();
        }
    }
}
```

锁屏的广播接收者在 SDK 中是一个 bug，如果在清单文件中配置是不能够生效的，必须通过代码配置才可以生效。另外，在停止该服务之前，必须反注册掉该广播接收者。

到此，程序锁功能全部完成。细心的读者应该发现了该功能同时使用到了 Android 的四大组件，所以，该功能也是对 Android 的四大组件的一个很好的复习，希望可以引起读者的重视。

需要掌握的知识点小结

（1）对数据库的优化——提取冗余数据。

（2）几种资源文件的复制方式：asset 目录下、res\raw 目录下、类加载器。

（3）使用 SQLiteDatabase.openDatabase() 打开已经存在的数据库。

（4）对数据库的增、删、改、查操作。

（6）使用 TelephoneyManager 监听系统电话的状态。

（7）服务的开启和关闭。

（8）使用 WindowsManager 模仿 Toast 显示一个自定义的界面。

（9）onTouch()、onClick()、onDoubleClick()的区别以及事件的消费。

（10）使用 ExpandableListView 实现 View 的分级显示。

（11）使用 ActivityManager 查看正在运行的服务。

（12）自定义 ContentProvider 实现数据的同步。

（13）使用广播接收者实现锁屏，以及代码注册广播的实现。

（14）使用 Service 充当看门狗，实现实时对任务栈栈顶的监视。

（15）借用 Activity 的 SingleInstance 启动模式来修复程序锁中的 bug。

第4章 通信卫士模块的设计

4.1 通信卫士的功能介绍与 UI 设计

功能需求：当进入"通信卫士"后，会将我们添加的黑名单号码用 ListView 展现出来。单击屏幕右上方的"添加黑名单号码"条目时，会弹出一个对话框让我们输入所要拦截的黑名单号码。同时也可以在该窗体中选择拦截模式（短信、电话），单击"确定"按钮后，ListView 中就会立即多出一条黑名单号码，如果在"设置中心"中开启了"来点黑名单设置"，拦截模式会立即生效。当我们长按其中的黑名单号码条目时，可以执行修改或删除黑名单号码的操作。这里没有"从手机联系人中获取号码"的选项，而是直接进入添加联系人的操作。在"手机防盗"中的设置向导里已经实现了从联系人列表中获取号码，感兴趣的读者可以将其添加进来。

实现思路：首先应当创建好数据，然后将数据一次性取出存入集合中，数据适配器从该集合中取出数据；当增加、删除、修改黑名单号码后，需要立即更改数据库和集合中的数据，紧接着让数据适配器重新适配数据；短信拦截可以采用广播接收者实现；电话拦截需要采用内容提供者和内容观察者（及时更新数据库中的数据）来操作手机系统中联系人的数据库。

将添加的黑名单号码存入数据库中，所以，需要创建这个数据库。将创建数据的对象 BlackNumberDBOpenHelper 放在包"com.guoshisp.mobilesafe.db"下，BlackNumberDBOpenHelper.java 代码如下：

```
package com.guoshisp.mobilesafe.db;
import android.content.Context;
import android.database.sqlite.SQLiteDatabase;
import android.database.sqlite.SQLiteDatabase.CursorFactory;
import android.database.sqlite.SQLiteOpenHelper;
public class BlackNumberDBOpenHelper extends SQLiteOpenHelper {
    public BlackNumberDBOpenHelper(Context context) {
        //参数一：上、下文对象；参数二：数据库名称；参数三：游标工厂对象，null 表示
          使用系统默认的；参数四：当前数据库的版本号
        super(context, "blacknumber.db", null, 1);
    }
    /**
     * 数据库第一次被创建时执行 oncreate()
     * 一般用于指定数据库的表结构
```

```java
     */
    @Override
    public void onCreate(SQLiteDatabase db) {
        //黑名单号码的表结构——_id、黑名单号码、拦截模式：0 表示电话拦截；1 表示短
        信拦截；2 表示全部拦截（电话与短信）
        db.execSQL("create table blacknumber (_id integer primary key 
                autoincrement, number varchar(20), mode integer)");
    }
    /**
     * 当数据库的版本号升级时调用的方法
     * 一般用于升级程序后，更新数据库的表结构
     */
    @Override
    public void onUpgrade(SQLiteDatabase db, int oldVersion, int newVersion) {
    }
}
```

下面来测试一下数据库是否被创建出来。我们需要在清单文件中配置一下测试环境：

（1）在</manifest>标签下添加一个子标签：

```xml
<instrumentation
    android:name="android.test.InstrumentationTestRunner"
    android:targetPackage="com.guoshisp.mobilesafe" />
```

（2）在</application>标签下添加一个子标签：

```xml
<uses-library android:name="android.test.runner" />
```

在包"com.guoshisp.mobilesafe.test"下创建一个测试类 TestCreatDB，TestCreatDB.java 代码如下：

```java
package com.guoshisp.mobilesafe.test;
import android.database.sqlite.SQLiteDatabase;
import android.test.AndroidTestCase;
import com.guoshisp.mobilesafe.db.BlackNumberDBOpenHelper;
public class TestCreatDB extends AndroidTestCase {
    public void testCreateDB() throws Exception {
        //测试数据库blacknumber.db是否会被创建
        BlackNumberDBOpenHelper helper = new BlackNumberDBOpenHelper(
                getContext());
        SQLiteDatabase db = helper.getWritableDatabase();
    }
}
```

测试运行：打开大纲视图 Outline 后，用鼠标左键单击一下 testCreateDB()方法→用鼠标右键单击该方法→Run As→Android Junit Test。等待数秒后，在 JUnit 控制台中出现一个绿条（如图 4-1 所示），这表示测试通过。进入 data\data\com.guoshisp.mobilesafe\databases 目录下可以看到数据库文件 blacknumber.db 已经生成，将其导出到桌面上，然后用 SQLite Expert 打开该数据库文件，然后打开 blacknumber 表，如图 4-2 所示（该表结构和我们创建时指定的表结构一样）。

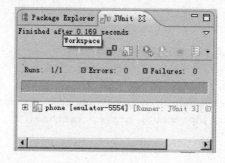

图 4-1

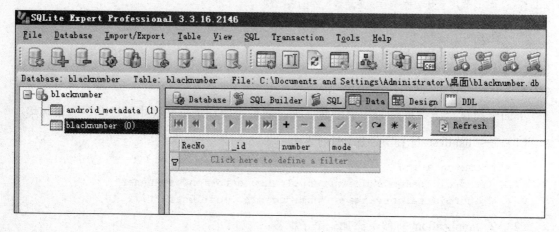

图 4-2

然后，在 SQLite Expert 工具中执行增、删、改、查的 SQL 语句。

➢ 添加两条黑名单记录（执行后的 blacknumber 表如图 4-3 所示，两条语句分开执行）：

```
insert into blacknumber (number,mode) values ('110','0')
insert into blacknumber (number,mode) values ('119','1')
```

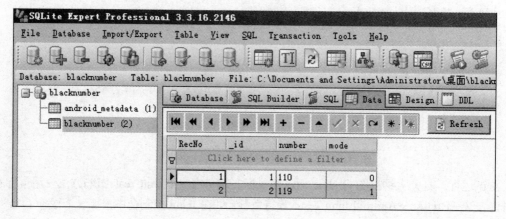

图 4-3

➢ 删除一条黑名单号码（执行后的 blacknumber 表如图 4-4 所示）：

```
delete from blacknumber where number='110'
```

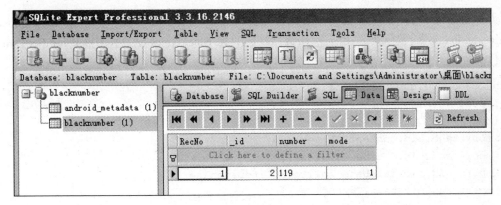

图 4-4

➤ 更改一条黑名单号码（执行后的 blacknumber 表如图 4-5 所示）：

```
update blacknumber set number='120', mode='2' where number='119'
```

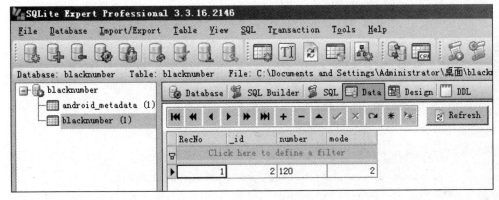

图 4-5

➤ 查找一条黑名单号码（执行后的 blacknumber 表如图 4-6 所示）：

```
select * from blacknumber where number ='120'
```

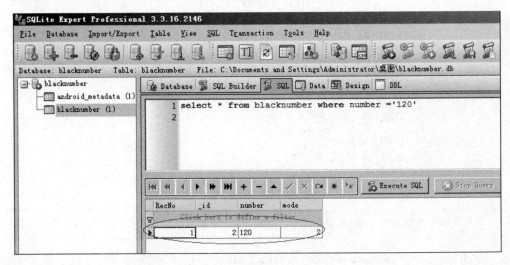

图 4-6

创建好数据库后,需要在代码中实现对数据库的增、删、改、查操作,在包"com. guoshisp. mobilesafe.db.dao"下创建一个用于操作数据库的对象 BlackNumberDao,BlackNumber Dao.java 中的代码如下:

```java
package com.guoshisp.mobilesafe.db.dao;
import java.util.ArrayList;
import java.util.List;
import android.content.Context;
import android.database.Cursor;
import android.database.sqlite.SQLiteDatabase;
import android.text.TextUtils;
import com.guoshisp.mobilesafe.db.BlackNumberDBOpenHelper;
import com.guoshisp.mobilesafe.domain.BlackNumber;
public class BlackNumberDao {
    private BlackNumberDBOpenHelper helper;
    public BlackNumberDao(Context context) {
        helper = new BlackNumberDBOpenHelper(context);
    }
    /**
     * 查找一条黑名单号码(其返回值用于判断数据库中是否存在该号码)
     */
    public boolean find(String number) {
        //默认情况下是没有该条数据
        boolean result = false;
        //打开数据库
        SQLiteDatabase db = helper.getReadableDatabase();
        if (db.isOpen()) {
            //执行查询语句后,返回一个结果集
            Cursor cursor = db.rawQuery(
                    "select * from blacknumber where number =?",
                    new String[] { number });
            //默认情况下,游标指针指向在第一条数据的上方
            if (cursor.moveToFirst()) {
                //返回true,说明数据库中已经存在该条数据
                result = true;
            }
            //关闭数据库
            cursor.close();
            db.close();
        }
        return result;
    }
    /**
     * 查找一条黑名单号码的拦截模式
     */
    public int findNumberMode(String number) {
```

```java
        //拦截模式只有3种：0代表拦截短信；1代表拦截电话；2代表拦截短信与电话。这
            里的默认值为-1,表示的是没有标记拦截模式
        int result = -1;
        SQLiteDatabase db = helper.getReadableDatabase();
        if (db.isOpen()) {
            //由于一条号码值对应一个拦截模式，所以，该结果集中只有一条数据
            Cursor cursor = db.rawQuery(
                    "select mode from blacknumber where number =?",
                    new String[] { number });
            if (cursor.moveToFirst()) {
                //获取第一条数据（也仅有一条数据）
                result = cursor.getInt(0);
            }
            cursor.close();
            db.close();
        }
        return result;
    }
    /**
     * 添加一条黑名单号码
     */
    public boolean add(String number, String mode) {
//首先判断数据库中是否已经存在该条数据,防止添加重复的数据显示到黑名单列表中
        if (find(number))
            //如果数据库中已经存在要添加的数据,直接停止掉该方法的执行
            return false;
        SQLiteDatabase db = helper.getWritableDatabase();
        if (db.isOpen()) {
            // 执行添加数据的SQL语句
            db.execSQL("insert into blacknumber (number,mode) values (?,?)",
                    new Object[] { number, mode });
            db.close();
        }
//如果代码能够执行到这一步,说明上面的添加操作也执行了,所以查询的返回值必定为true
        return find(number);
    }
    /**
     * 删除一条黑名单号码
     */
    public void delete(String number) {
        SQLiteDatabase db = helper.getWritableDatabase();
        if (db.isOpen()) {
            //执行删除操作
            db.execSQL("delete from blacknumber where number=?",
                    new Object[] { number });
            db.close();
```

```java
        }
    }
    /**
     * 更改黑名单号码
     *
     * @param oldnumber
     * 旧的的电话号码
     * @param newnumber
     * 新的号码,可以留空
     * @param mode
     * 新的模式
     */
    public void update(String oldnumber, String newnumber, String mode) {

        SQLiteDatabase db = helper.getWritableDatabase();
        if (db.isOpen()) {
            if (TextUtils.isEmpty(newnumber)) {
//如果新的号码为空,则说明用户并没有修改该号码(ListView 中的 item 设置有删除功能)
                newnumber = oldnumber;
            }
            //执行更新操作
            db.execSQL(
                    "update blacknumber set number=?, mode=? where number=?",
                    new Object[] { newnumber, mode, oldnumber });
            db.close();
        }
    }
    /**
     * 查找全部的黑名单号码
     *
     * @return
     */
    public List<BlackNumber> findAll() {
        //定义好要返回的对象
        List<BlackNumber> numbers = new ArrayList<BlackNumber>();
        SQLiteDatabase db = helper.getReadableDatabase();
        if (db.isOpen()) {
            //查询 blacknumber 表中的所有号码
            Cursor cursor = db.rawQuery("select number,mode from blacknumber",
                    null);
            //循环遍历结果集,将每个结果集封装后添加到集合中
            while (cursor.moveToNext()) {
                BlackNumber blackNumber = new BlackNumber();
                blackNumber.setNumber(cursor.getString(0));
                blackNumber.setMode(cursor.getInt(1));
                numbers.add(blackNumber);
```

```
            blackNumber = null;
        }
        cursor.close();
        db.close();
    }
    return numbers;
    }
}
```

下面，我们来测试一下该对象能否正确地执行对数据库的增、删、改、查操作。在包"com.guoshisp.mobilesafe.test"下面新建一个测试类 TestBlackNumberDao，TestBlackNumberDao.java 中的代码如下：

```
package com.guoshisp.mobilesafe.test;
import java.util.List;
import java.util.Random;
import android.test.AndroidTestCase;
import com.guoshisp.mobilesafe.db.dao.BlackNumberDao;
import com.guoshisp.mobilesafe.domain.BlackNumber;
public class TestBlackNumberDao extends AndroidTestCase {
    //向数据库中添加50条数据
    public void testAdd() throws Exception {
        BlackNumberDao dao = new BlackNumberDao(getContext());
        //第一个要被添加的号码
        int number = 100000;
        Random random = new Random();
        for (int i = 0; i < 50; i++) {
            int result = (number+i);
            //执行添加操作。random.nextInt(3)表示的随机数为0、1、2
            dao.add(result+"", random.nextInt(3)+"");
        }
    }
    //更新数据库中的数据
    public void testUpdate() throws Exception {
        BlackNumberDao dao = new BlackNumberDao(getContext());
        dao.update("100000", "999999", "2");
    }
    //删除数据库中的数据
    public void testDelete() throws Exception {
        BlackNumberDao dao = new BlackNumberDao(getContext());
        dao.delete("999999");
    }
    //查询数据库表中的所有的号码
    public void testFindAll() throws Exception {
        BlackNumberDao dao = new BlackNumberDao(getContext());
        List<BlackNumber> numbers = dao.findAll();
```

```
            System.out.println(numbers.size());
        }
    }
```

按顺序分别测试上面的四个方法,对于前三个方法:每测试一个方法后,我们将数据库导出,并用 SQLite Expert 工具打开,第四个方法的执行结果可以在 Log 日志中看到,对应的图分别为图 4-7、图 4-8、图 4-9、图 4-10。

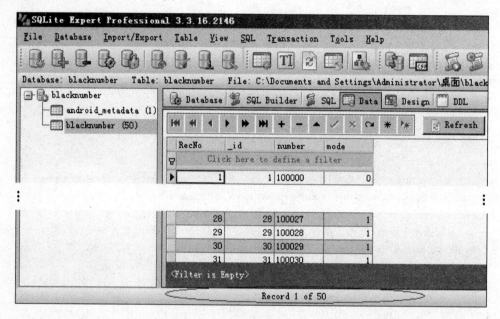

图 4-7

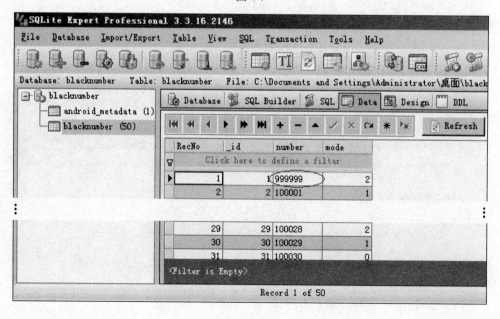

图 4-8

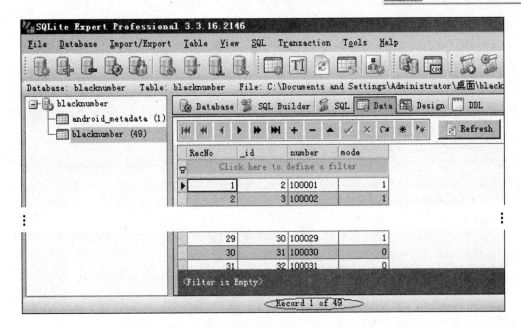

图 4-9

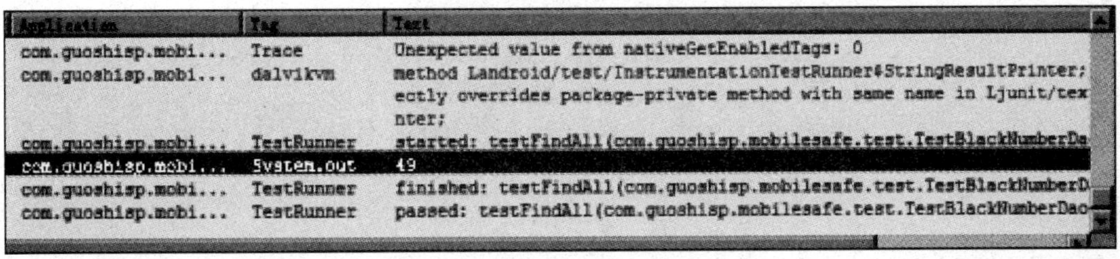

图 4-10

准备好数据后,接下来需要开发出"通信卫士"的界面,并将数据显示在界面上。"通信卫士"所对应的 Activity 为 CallSmsSafeActivity,CallSmsSafeActivity.java 对应的业务代码如下:

```
package com.guoshisp.mobilesafe;
import java.util.List;
import android.app.Activity;
import android.os.Bundle;
import android.os.Handler;
import android.os.Message;
import android.util.Log;
import android.view.View;
import android.view.ViewGroup;
import android.widget.BaseAdapter;
import android.widget.LinearLayout;
import android.widget.ListView;
import android.widget.TextView;
import com.guoshisp.mobilesafe.db.dao.BlackNumberDao;
```

```java
import com.guoshisp.mobilesafe.domain.BlackNumber;
public class CallSmsSafeActivity extends Activity {
    protected static final int LOAD_DATA_FINISH = 40;
    public static final String TAG = "CallSmsSafeActivity";
    //用于展现出所有的黑名单号码
    private ListView lv_call_sms_safe;
    //操作黑名单号码数据库的对象
    private BlackNumberDao dao;
    //将黑名单号码从数据库中一次性取出存入缓存集合中（避免在适配器中频繁地操作数据库）
    private List<BlackNumber> blacknumbers;
    //显示黑名单号码的适配器对象
    private BlackNumberAdapter adpater;
    //ProgressBar 控件的父控件，用于控制子控件的显示（包括了 ProgressBar）
    private LinearLayout ll_call_sms_safe_loading;
    //用于接收子线程发送过来的消息，实现 UI 的更新
    private Handler handler = new Handler() {
        public void handleMessage(android.os.Message msg) {
            switch (msg.what) {
            case LOAD_DATA_FINISH://从数据库中加载黑名单号码完成
                //将进度条及"正在加载数据..."隐藏
                ll_call_sms_safe_loading.setVisibility(View.INVISIBLE);
                //为 lv_call_sms_safe 设置适配器
                adpater = new BlackNumberAdapter();
                lv_call_sms_safe.setAdapter(adpater);
                break;
            }
        };
    };
    @Override
    protected void onCreate(Bundle savedInstanceState) {
        super.onCreate(savedInstanceState);
        setContentView(R.layout.call_sms_safe);
        ll_call_sms_safe_loading = (LinearLayout) findViewById(R.id.ll_
                            call_sms_safe_loading);
        dao = new BlackNumberDao(this);
        lv_call_sms_safe = (ListView) findViewById(R.id.lv_call_sms_safe);
        //ll_call_sms_safe_loading 控件中的所有子控件设置为可见（ProgressBar
        //  和"正在加载数据…"）
        ll_call_sms_safe_loading.setVisibility(View.VISIBLE);
        //一次性获取数据库中的所有数据的操作是一个比较耗时的操作，建议在子线程中完成
        new Thread() {
            public void run() {
                blacknumbers = dao.findAll();
                //通知主线程更新界面
                Message msg = Message.obtain();
                msg.what = LOAD_DATA_FINISH;
```

```java
                handler.sendMessage(msg);
            };
        }.start();
    }
    /**
     * 为黑名单号码中的 lv_call_sms_safe 中的 Item 适配数据
     * @author Administrator
     *
     */
    private class BlackNumberAdapter extends BaseAdapter {
        //获取 Item 的数目
        public int getCount() {
            return blacknumbers.size();
        }
        //获取 Item 的对象
        public Object getItem(int position) {
            return blacknumbers.get(position);
        }
        //获取 Item 对应的 id
        public long getItemId(int position) {
            return position;
        }
        //在屏幕上,每显示一个 Item 就调用一次该方法
        public View getView(int position, View convertView, ViewGroup
                            parent) {
            View view;
            ViewHolder holder;
            //复用历史缓存的 View 对象
            if (convertView == null) {
                Log.i(TAG, "创建新的 view 对象");
                //将 Item 转成 View 对象
                view = View.inflate(getApplicationContext(),
                        R.layout.call_sms_item, null);
                holder = new ViewHolder();
                holder.tv_number = (TextView) view
                        .findViewById(R.id.tv_callsms_item_number);
                holder.tv_mode = (TextView) view
                        .findViewById(R.id.tv_callsms_item_mode);
                view.setTag(holder);//把控件 id 的引用存放在 view 对象中
            } else {
                view = convertView;
                Log.i(TAG, "使用历史缓存的 view 对象");
                holder = (ViewHolder) view.getTag();
            }
            //为 Item 设置拦截模式
            BlackNumber blacknumber = blacknumbers.get(position);
```

```
                holder.tv_number.setText(blacknumber.getNumber());
                int mode = blacknumber.getMode();
                if (mode == 0) {
                    holder.tv_mode.setText("电话拦截");
                } else if (mode == 1) {
                    holder.tv_mode.setText("短信拦截");
                } else {
                    holder.tv_mode.setText("全部拦截");
                }
                return view;
            }
        }
        //将Item中的控件使用static修饰，被static修饰的类的字节码在JVM中只会存在一
          份，tv_number与tv_mode在栈中也会只存在一份
        private static class ViewHolder {
            TextView tv_number;
            TextView tv_mode;
        }
    }
```

代码解析：

（1）ll_call_sms_safe_loading.setVisibility(View.INVISIBLE)

ll_call_sms_safe_loading所对应的控件中包含ProgressBar和TextView（显示"正在加载数据…"），执行该代码后，ll_call_sms_safe_loading控件中的所有控件的显示都被隐藏。

（2）blacknumbers = dao.findAll()

将数据库中的数据一次性查出并存放到一个集合中。在适配数据时，可以直接从缓存中（集合）取出数据，从而避免了频繁地操作数据库。如果频繁地操作数据库，会导致很多GC信息产生，同时在滚动ListView时还会出现卡屏的现象。

（3）view = convertView

convertView是缓存中的View，这里是复用了缓存中的view对象，从而避免了通过布局填充器得到view对象（执行布局填充器代码是比较耗时的），这也是对ListView优化的有效方式。

该界面在layout下的布局文件call_sms_safe.xml文件内容如下：

```xml
<?xml version="1.0" encoding="utf-8"?>
<LinearLayout xmlns:android="http://schemas.android.com/apk/res/android"
    android:layout_width="fill_parent"
    android:layout_height="fill_parent"
    android:orientation="vertical" >
    <TextView
        android:layout_width="wrap_content"
        android:layout_height="wrap_content"
        android:layout_gravity="center_horizontal"
```

```xml
        android:text="通信卫士"
        android:textColor="#66ff00"
        android:textSize="28sp" />
    <View
        android:layout_width="fill_parent"
        android:layout_height="1dip"
        android:layout_marginTop="5dip"
        android:background="@drawable/devide_line" />
    <RelativeLayout
        android:layout_width="fill_parent"
        android:layout_height="fill_parent" >
        <Button
            android:layout_width="wrap_content"
            android:layout_height="wrap_content"
            android:layout_alignParentRight="true"
            android:background="@drawable/bt_selector"
            android:onClick="addBlackNumber"
            android:text="添加黑名单号码" />
        <LinearLayout
            android:id="@+id/ll_call_sms_safe_loading"
            android:layout_width="fill_parent"
            android:layout_height="fill_parent"
            android:gravity="center"
            android:orientation="vertical"
            android:visibility="invisible" >
            <ProgressBar
                android:layout_width="wrap_content"
                android:layout_height="wrap_content" />
            <TextView
                android:layout_width="wrap_content"
                android:layout_height="wrap_content"
                android:text="正在加载数据..." />
        </LinearLayout>
        <ListView
            android:id="@+id/lv_call_sms_safe"
            android:layout_width="fill_parent"
            android:layout_height="fill_parent"
            android:layout_marginTop="5dip" >
        </ListView>
    </RelativeLayout>
</LinearLayout>
```

通过该布局文件可以看出：我们为 Button 设置了一个背景资源，该背景资源位于 drawable 目录下，bt_selector.xml 文件内容如下：

```xml
<?xml version="1.0" encoding="utf-8"?>
```

```xml
<selector xmlns:android="http://schemas.android.com/apk/res/android">
    <item android:state_pressed="true"
        android:drawable="@drawable/bt_selected" /> <!-- pressed -->
    <item android:state_focused="true"
        android:drawable="@drawable/bt_selected" /> <!-- focused -->
    <item android:drawable="@drawable/bt_normal" /> <!-- default -->
</selector>
```

在该背景资源中定义了 Button 按钮被按下、获取焦点、正常时的状态（图形和颜色），其中，当按钮被按下和获取到焦点时的状态同为 bt_selected.xml，正常状态下为 bt_normal.xml，资源都存放在 drawable 目录下，两者都定义了一个矩形的图形资源，区别在于颜色的不同。bt_selected.xml 文件内容如下：

```xml
<?xml version="1.0" encoding="utf-8"?>
<shape xmlns:android="http://schemas.android.com/apk/res/android"
    android:shape="rectangle" >
    <corners android:radius="3dip" />
    <solid android:color="#66ff02" />
    <padding
        android:bottom="3dip"
        android:left="3dip"
        android:right="3dip"
        android:top="3dip" />
</shape>
```

bt_normal.xml 文件内容如下：

```xml
<?xml version="1.0" encoding="utf-8"?>
<shape xmlns:android="http://schemas.android.com/apk/res/android"
    android:shape="rectangle" >
    <corners android:radius="3dip" />
    <solid android:color="#55ff00" />
    <padding
        android:bottom="3dip"
        android:left="3dip"
        android:right="3dip"
        android:top="3dip" />
</shape>
```

在 ListView 显示数据时，Item 对应的布局文件存放在 layout 目录下，call_sms_item.xml 文件内容如下：

```xml
<?xml version="1.0" encoding="utf-8"?>
<RelativeLayout xmlns:android="http://schemas.android.com/apk/res/android"
    android:layout_width="match_parent"
    android:layout_height="wrap_content" >
    <TextView
        android:id="@+id/tv_callsms_item_number"
```

```xml
        style="@style/text_content_style"
        android:layout_width="wrap_content"
        android:text="电话号码" />
    <TextView
        android:id="@+id/tv_callsms_item_mode"
        style="@style/text_content_style"
        android:layout_width="wrap_content"
        android:layout_alignParentBottom="true"
        android:layout_alignParentRight="true"
        android:text="拦截模式"
        android:textSize="12sp" />
</RelativeLayout>
```

最后还需要在主界面中为"通信卫士"设置单击事件，同时在清单文件中做好组件信息的配置，MainActivity.java 中的业务代码如下：

```java
package com.guoshisp.mobilesafe;
import android.app.Activity;
import android.content.Intent;
import android.os.Bundle;
import android.view.View;
import android.widget.AdapterView;
import android.widget.AdapterView.OnItemClickListener;
import android.widget.GridView;
import com.guoshisp.mobilesafe.adapter.MainAdapter;
public class MainActivity extends Activity {
    //显示主界面中九大模块的GridView
    private GridView gv_main;
    @Override
    protected void onCreate(Bundle savedInstanceState) {
        super.onCreate(savedInstanceState);
        setContentView(R.layout.main);
        gv_main = (GridView) findViewById(R.id.gv_main);
    //为gv_main对象设置一个适配器，该适配器的作用是用于为每个item填充对应的数据
        gv_main.setAdapter(new MainAdapter(this));
    //为GridView对象中的item设置单击时的监听事件
        gv_main.setOnItemClickListener(new OnItemClickListener() {
            //参数一：item的父控件，也就是GridView；参数二：当前单击的item
            //参数三：当前单击的item在GridView中的位置
            //参数四：id的值为单击了GridView的哪一项对应的数值，单击了GridView
                第9项，那id就等于8
            public void onItemClick(AdapterView<?> parent, View view,
                    int position, long id) {
                switch (position) {
                case 0: //手机防盗
                    //跳转到"手机防盗"对应的Activity界面
```

```
            Intent lostprotectedIntent = new Intent(MainActivity.
                this,LostProtectedActivity.class);
            startActivity(lostprotectedIntent);
            break;
        case 1: //通信卫士
            Intent callSmsIntent = new Intent(MainActivity.this,
                        CallSmsSafeActivity.class);
            startActivity(callSmsIntent);
            break;
        case 7://高级工具
            Intent atoolsIntent = new Intent(MainActivity.this,
                        AtoolsActivity.class);
            startActivity(atoolsIntent);
            break;
        case 8://设置中心
            //跳转到"设置中心"对应的Activity界面
            Intent settingIntent = new Intent(MainActivity.this,
                        SettingCenterActivity.class);
            startActivity(settingIntent);
            break;
        }
    }
});
    }
}
```

测试运行：单击"通信卫士"时加载数据的界面效果如图 4-11 所示，加载数据完毕后的界面效果如图 4-12 所示；滑动 ListView 时的界面效果如图 4-13 所示。

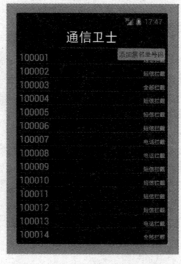

图 4-11　　　　　　　　　图 4-12　　　　　　　　　图 4-13

在图 4-13 中我们发现：当滑动 ListView 时，添加号码的 Button 会被遮挡住。原因在于：Button 和 ListView 控件同在 RelativeLayout 控件里面，且 Button 控件位于 ListView 控件的上面，而 RelativeLayout 中的子控件在被渲染时是按照自上而下的顺序，后被渲染的控件会覆盖已经渲染的控件。所以，我们将 Button 控件作为 RelativeLayout 控件的最后一个子控件即可解决该 bug。

4.2　黑名单号码的添加与修改

需求：在黑名单号码列表中单击"添加黑名单号码"按钮后，应当弹出一个对话框，在该对话框中实现对黑名单号码的添加，以及设置该号码的拦截模式。同时，要避免列表中有重复的号码，添加成功后，立即将数据更新到当前列表中显示。当长按一个 Item 时，会弹出一个窗体，该窗体中的值存在两个条目（分别显示"删除黑名单号码"和"更改黑名单号码"），当选中第一个条目后，该 Item 对应的黑名单号码会立刻被删除；当选中第二个条目时，会弹出一个窗体对话框（该对话框与添加黑名单对话框只在标题上不同），并且将 Item 对应的号码和拦截模式显示到对话框上；当单击"确定"按钮时，会执行对数据库的查询操作，检查是否已经存在该号码，避免修改为已经存在的黑名单号码。当修改完毕后，立即在当前界面上生效。

实现思路：添加黑名单的对话框是以窗体的样式显示出来的，在创建窗体的过程中，将一个布局文件转换成 View 后添加到窗体中；通过对电话拦截和短信拦截前面的 Checkbox 是否被勾选，来判断出所要存入数据库的拦截模式的值；每当添加一条数据时，先查找数据库中是否存在该数据，这样可以避免重复的数据被添加进数据库中；当添加数据成功后，通知适配器来更新列表中的数据。

修改黑名单号码：可以通过 Contextual Menus 来实现长按 Item 弹出一个对话框。实现该功能只需要三步。

（1）为 ListView 注册一个上、下文菜单：registerForContextMenu(View view)。

（2）重写创建上、下文菜单的方法：onCreateContextMenu(ContextMenu menu, View v, ContextMenuInfo menuInfo)。

（3）响应上、下文菜单的单击事件：onContextItemSelected(MenuItem item)。

单击删除时，执行的操作是将该号码从数据库中删除后，再从集合中删除（适配数据时是从集合中取出的）。

代码部分见在线资源包中代码文本部分 4.3.doc。

在 inflater.inflate(R.menu.call_sms_safe_menu, menu)中，call_sms_safe_menu.xml 就是长按 Item 时要显示的一个小窗体布局文件，该文件位于 res\menu 目录下，文件内容如下：

```xml
<?xml version="1.0" encoding="utf-8"?>
<menu xmlns:android="http://schemas.android.com/apk/res/android" >
    <item
        android:id="@+id/item_delete"
        android:title="删除黑名单号码">
    </item>
    <item
        android:id="@+id/item_update"
        android:title="更改黑名单号码">
    </item>
</menu>
```

说明：在showBlackNumberDialog(final int flag, final int position)方法中，通过flag来判断当前是添加黑名单号码还是修改黑名单号码，根据position来判断Item的位置。

测试运行：单击界面中的"添加黑名单号码"按钮时，弹出一个窗体，我们输入黑名单号码为666666，拦截模式选择电话拦截，界面效果如图4-14所示。单击"确定"按钮后，该黑名单号码立即显示在界面上，如图4-15所示。长按666666所对应的Item时弹出一个窗体，如图4-16所示。在该窗体中单击"修改黑名单号码"时，弹出对话框如图4-14所示（号码和拦截模式自动回显），将其修改为888888，并将拦截模式修改为电话拦截和短信拦截，单击"确定"按钮，立即在界面中生效，如图4-17所示。当我们长按888888条目后，选择"删除黑名单号码"后，该号码立即在该列表中被移除，同图4-13一样。

图4-14

图4-15

图 4-16　　　　　　　　　图 4-17

以上实现了对黑名单的增、删、该操作。下面需要实现黑名单号码相应的拦截功能。

4.3　黑名单号码对短信和电话的拦截

首先实现短信黑名单的拦截，在实现"手机防盗"功能时，通过广播在 SmsReceiver.java 中实现安全号码发送指令的操作。同样，我们也可以将黑名单号码短信拦截的逻辑在此实现，SmsReceiver.java 业务代码如下：

```java
package com.guoshisp.mobilesafe.receiver;
import android.app.admin.DevicePolicyManager;
import android.content.BroadcastReceiver;
import android.content.ComponentName;
import android.content.Context;
import android.content.Intent;
import android.content.SharedPreferences;
import android.media.MediaPlayer;
import android.telephony.SmsManager;
import android.telephony.SmsMessage;
import android.text.TextUtils;
import android.util.Log;
import com.guoshisp.mobilesafe.R;
import com.guoshisp.mobilesafe.db.dao.BlackNumberDao;
import com.guoshisp.mobilesafe.engine.GPSInfoProvider;
public class SmsReceiver extends BroadcastReceiver {
    private static final String TAG = "SmsReceiver";
    private SharedPreferences sp;
```

```java
private BlackNumberDao dao;
@Override
public void onReceive(Context context, Intent intent) {
    Log.i(TAG,"短信到来了");
    dao = new BlackNumberDao(context);
    sp = context.getSharedPreferences("config", Context.MODE_PRIVATE);
    String safenumber = sp.getString("safemuber", "");
//获取短信中的内容。系统接收到一个信息广播时,会将接收到的信息存放到pdus数组中
    Object[] objs = (Object[]) intent.getExtras().get("pdus");
    //获取手机设备管理器
    DevicePolicyManager dm = (DevicePolicyManager) context.getSystemService
                    (Context.DEVICE_POLICY_SERVICE);
    //创建一个与MyAdmin相关联的组件
    ComponentName mAdminName = new ComponentName(context, MyAdmin.class);
    //遍历出信息中的所有内容
    for(Object obj : objs){
    SmsMessage smsMessage = SmsMessage.createFromPdu((byte[]) obj);
        //获取发件人的号码
        String sender = smsMessage.getOriginatingAddress();
        //判断短信号码是否是黑名单号码与短信拦截
        int result = dao.findNumberMode(sender);
        if(result==1||result==2){//判断该黑名单号码是否需要拦截短信
            Log.i(TAG,"拦截黑名单短信");
            abortBroadcast();
        }
        //获取短信信息内容
        String body = smsMessage.getMessageBody();
        if("#*location*#".equals(body)){
            Log.i(TAG,"发送位置信息");
            //获取上次的位置
            String lastlocation = GPSInfoProvider.getInstance(context).
                            getLocation();
            if(!TextUtils.isEmpty(lastlocation)){
                //得到信息管理器
                SmsManager smsManager = SmsManager.getDefault();
                //向安全号码发送当前的位置信息
            smsManager.sendTextMessage(safenumber, null, lastlocation,
                            null, null);
            }
            abortBroadcast();
        }else if("#*alarm*#".equals(body)){
            Log.i(TAG,"播放报警音乐");
            //得到音频播放器
            MediaPlayer player = MediaPlayer.create(context, R.raw.ylzs);
                            //res\raw\ylzs.mp3
            //即使手机是静音模式也有音乐的声音
            player.setVolume(1.0f, 1.0f);
```

```
            //开始播放音乐
            player.start();
            //终止掉发送过来的信息，在本地查看不到该信息
            abortBroadcast();
        }else if("#*wipedata*#".equals(body)){
            Log.i(TAG,"清除数据");
            //判断设备的管理员权限是否被激活。只有被激活后，才可以执行锁屏、清
              除数据恢复至出厂设置（模拟器不支持该操作）等操作
            if(dm.isAdminActive(mAdminName)){
                dm.wipeData(0);//清除数据恢复至出厂设置。手机会自动重启
            }
            abortBroadcast();
        }else if("#*lockscreen*#".equals(body)){
            Log.i(TAG,"远程锁屏");
            if(dm.isAdminActive(mAdminName)){
            dm.resetPassword("123", 0);//屏幕解锁时需要的解锁密码123
                dm.lockNow();
            }
            abortBroadcast();
        }
    }
}
```

短信拦截的实现逻辑：直接查询该短信号码在黑名单数据库中的拦截模式。

测试运行：使用 Eclipse 的 Emulator Control 来模拟 100001 号码进行短信的发送，单击发送后，Log 日志如图 4-18 所示。

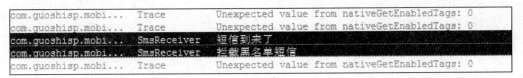

图 4-18

实现了短信的拦截后，我们来实现黑名单号码的拦截。

4.4 黑名单号码对电话的拦截

需求：当黑名单号码呼入时，手机自动将电话挂断，并且不会在通话记录中产生记录。同时，在"设置中心"中设置来电黑名单号码的开启、关闭设置。

实现思路：使用服务监听电话的状态，当电话为铃响状态时，判断呼入的号码是否属于黑名单号码，如果是，需要判断一下拦截模式，如果是电话拦截，使用系统提供的服务来将电话

直接挂断。挂断电话的操作可以通过 PackageManager 对象来实现，但在 Android1.5 以后，该方法没有被暴露出来，需要通过 aidl 来实现。删除来电记录是通过操作手机系统中的联系人的数据库来实现的。

首先，将挂断电话需要的两个 aidl 文件导入 src 目录下，NeighboringCellInfo.aidl 文件位于包"android.telephony"下，NeighboringCellInfo.aidl 文件内容如下：

```
package android.telephony;
parcelable NeighboringCellInfo;
```

另一个 aidl 文件 ITelephony.aidl 位于包"com.android.internal.telephony"下，ITelephony.aidl 文件内容如下：

```
package com.android.internal.telephony;
import android.os.Bundle;
import java.util.List;
import android.telephony.NeighboringCellInfo;
interface ITelephony {
    void dial(String number);
    void call(String number);
    boolean showCallScreen();
    boolean showCallScreenWithDialpad(boolean showDialpad);
    boolean endCall();
    void answerRingingCall();
    void silenceRinger();
    boolean isOffhook();
    boolean isRinging();
    boolean isIdle();
    boolean isRadioOn();
    boolean isSimPinEnabled();
    void cancelMissedCallsNotification();
    boolean supplyPin(String pin);
    boolean handlePinMmi(String dialString);
    void toggleRadioOnOff();
    boolean setRadio(boolean turnOn);
    void updateServiceLocation();
    void enableLocationUpdates();
    void disableLocationUpdates();
    int enableApnType(String type);
    int disableApnType(String type);
    boolean enableDataConnectivity();
    boolean disableDataConnectivity();
    boolean isDataConnectivityPossible();
    Bundle getCellLocation();
    List<NeighboringCellInfo> getNeighboringCellInfo();
    int getCallState();
    int getDataActivity();
    int getDataState();
```

```
    int getActivePhoneType();
    int getCdmaEriIconIndex();
    int getCdmaEriIconMode();
    String getCdmaEriText();
    boolean getCdmaNeedsProvisioning();
    int getVoiceMessageCount();
    int getNetworkType();
    boolean hasIccCard();
}
```

接着,我们先实现电话拦截的基本功能,该类位于包 "com.guoshisp.mobilesafe. service" 下,CallFirewallService.java 中的业务方法如下:

```java
package com.guoshisp.mobilesafe.service;
import java.lang.reflect.Method;
import android.app.Service;
import android.content.Intent;
import android.os.IBinder;
import android.telephony.PhoneStateListener;
import android.telephony.TelephonyManager;
import android.util.Log;
import com.android.internal.telephony.ITelephony;
import com.guoshisp.mobilesafe.db.dao.BlackNumberDao;
public class CallFirewallService extends Service {
    public static final String TAG = "CallFirewallService";
    private TelephonyManager tm;
    private MyPhoneListener listener;
    private BlackNumberDao dao;
    @Override
    public IBinder onBind(Intent intent) {
        return null;
    }
    /**
     * 当服务第一次被创建时调用
     */
    @Override
    public void onCreate() {
        super.onCreate();
        dao = new BlackNumberDao(this);
        //注册系统的电话状态改变的监听器
        listener = new MyPhoneListener();
        tm = (TelephonyManager) getSystemService(TELEPHONY_SERVICE);
        //系统的电话服务监听了电话状态的变化
        tm.listen(listener, PhoneStateListener.LISTEN_CALL_STATE);
    }
    private class MyPhoneListener extends PhoneStateListener {
        @Override
        public void onCallStateChanged(int state, String incomingNumber) {
```

```java
            switch (state) {
            case TelephonyManager.CALL_STATE_RINGING://手机铃声正在响
                //判断 incomingNumber 是否是黑名单号码
                int mode = dao.findNumberMode(incomingNumber);
                if (mode == 0 || mode == 2) {
                    //黑名单号码
                    Log.i(TAG, "挂断电话");
                    //挂断电话
                    endcall(incomingNumber);
                }
                break;
            case TelephonyManager.CALL_STATE_IDLE: //手机的空闲状态
                break;
            case TelephonyManager.CALL_STATE_OFFHOOK://手机接通通话的状态
                break;
            }
            super.onCallStateChanged(state, incomingNumber);
        }
    }
    /**
     * 取消电话状态的监听
     */
    @Override
    public void onDestroy() {
        super.onDestroy();
        tm.listen(listener, PhoneStateListener.LISTEN_NONE);
        listener = null;
    }
    /**
     * 挂断电话
     * 需要复制两个 aidl 文件
     * 添加权限<uses-permission android:name="android.permission.CALL_PHONE"/>
     * @param incomingNumber
     */
    public void endcall(String incomingNumber) {
        try {
            //使用反射获取系统的 service 方法
            Method method = Class.forName("android.os.ServiceManager")
                    .getMethod("getService", String.class);
            IBinder binder = (IBinder) method.invoke(null,
                    new Object[] { TELEPHONY_SERVICE });
            //通过 aidl 实现方法的调用
            ITelephony telephony = ITelephony.Stub.asInterface(binder);
            telephony.endCall();//该方法是一个异步方法，会新开启一个线程将呼入的
                                //号码存入数据库中
        } catch (Exception e) {
            e.printStackTrace();
```

            }
        }

然后，在"设置中心"中添加"来电黑名单设置"是否开启的条目，SettingCenterActivity.java 业务代码及 layout 目录下的布局文件 Setting_center.xml 代码见在线资源包中代码文本部分 4.4.doc。

layout 目录下的布局文件 setting_center.xml 内容如下：

```xml
<?xml version="1.0" encoding="utf-8"?>
<LinearLayout xmlns:android="http://schemas.android.com/apk/res/android"
    android:layout_width="fill_parent"
    android:layout_height="fill_parent"
    android:orientation="vertical" >
    <TextView
        android:layout_width="wrap_content"
        android:layout_height="wrap_content"
        android:layout_gravity="center_horizontal"
        android:text="设置中心"
        android:textColor="#66ff00"
        android:textSize="28sp" />
    <View
        android:layout_width="fill_parent"
        android:layout_height="1dip"
        android:layout_marginTop="5dip"
        android:background="@drawable/devide_line" />
    <RelativeLayout
        android:layout_width="fill_parent"
        android:layout_height="wrap_content"
        android:layout_marginTop="8dip" >
        <TextView
            android:id="@+id/tv_setting_autoupdate_text"
            android:layout_width="wrap_content"
            android:layout_height="wrap_content"
            android:text="自动更新设置"
            android:textSize="26sp"
            android:textStyle="bold" >
```

测试运行：开启设置中心中的来电黑名单设置。我们先模拟一下使用电话的三种情况：模拟呼入 100008 号码（黑名单电话拦截，Log 日志如图 4-19 所示）；模拟外拨 111111 号码，然后挂断；模拟呼入 555555 不接听，直接挂断。手机系统中的联系人数据库位于 data\data\com.android.providers.contacts\databases 目录下的 contacts2.db 文件中。此时，将数据库导入后，使用 SQLite Expert 打开该数据库，单击 calls 表查看数据，如图 4-20 所示。

com.guoshisp.mobi...	Trace	Unexpected value from nativeGetEnabledTags: 0
com.guoshisp.mobi...	Trace	Unexpected value from nativeGetEnabledTags: 0
com.guoshisp.mobi...	CallFirew...	挂断电话
com.guoshisp.mobi...	Trace	Unexpected value from nativeGetEnabledTags: 0
com.guoshisp.mobi...	Trace	Unexpected value from nativeGetEnabledTags: 0

图 4-19

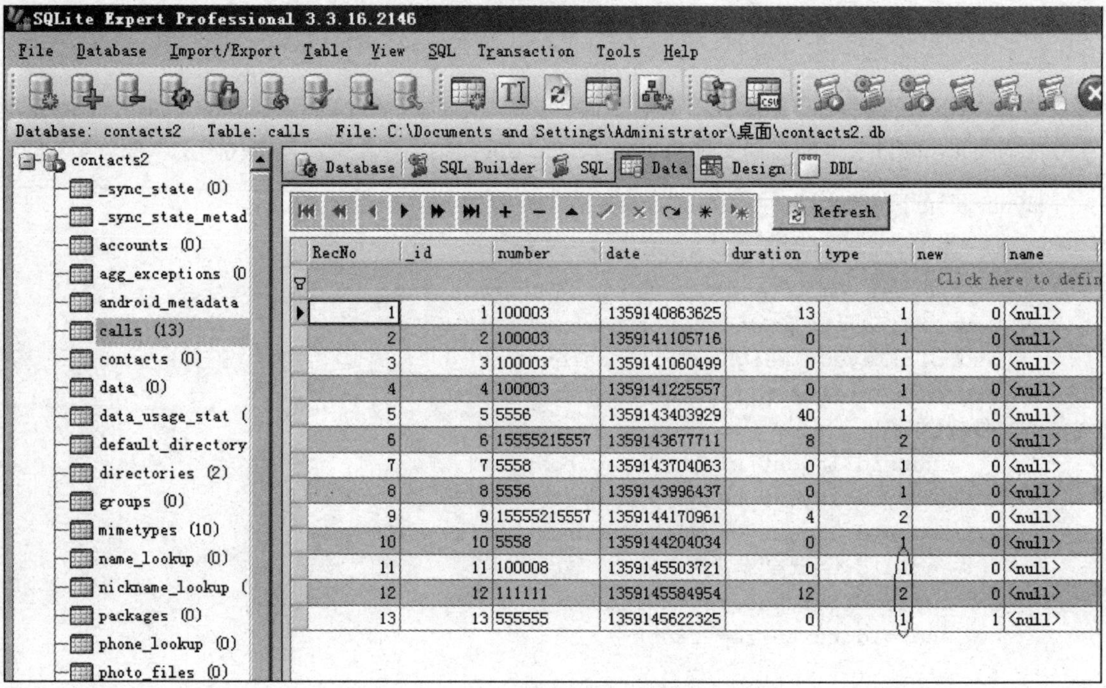

图 4-20

由图 4-19 可以发现，呼入号码（包含呼入未接）的 type 都为 1，而呼出的 type 为 2。与 Android2.3 环境下的不同：呼入的 type 为 1（包含黑名单拦截挂断），呼出的 type 为 2，未接的 type 为 3。我们可以根据 type 的不同，来删除联系人中的号码。

由图 4-19 可知，100008 黑名单电话拦截模式的号码确实已经被拦截下来，但是打开通话记录时，却发现存在通话记录，如图 4-21 所示。

图 4-21

前面已经分析了手机系统中联系人所对应的数据库,我们可以这样实现:每当拦截一次号码,就到数据库中将该号码删除掉。此时,CallFirewallService.java 对应的业务代码如下:

```java
package com.guoshisp.mobilesafe.service;
import java.lang.reflect.Method;
import android.app.Service;
import android.content.Intent;
import android.database.Cursor;
import android.net.Uri;
import android.os.IBinder;
import android.telephony.PhoneStateListener;
import android.telephony.TelephonyManager;
import android.util.Log;
import com.android.internal.telephony.ITelephony;
import com.guoshisp.mobilesafe.db.dao.BlackNumberDao;
public class CallFirewallService extends Service {
    public static final String TAG = "CallFirewallService";
    private TelephonyManager tm;
    private MyPhoneListener listener;
    private BlackNumberDao dao;
    @Override
    public IBinder onBind(Intent intent) {
        return null;
    }
    /**
     * 当服务第一次被创建时调用
     */
    @Override
    public void onCreate() {
        super.onCreate();
        dao = new BlackNumberDao(this);
        //注册系统的电话状态改变的监听器
        listener = new MyPhoneListener();
        tm = (TelephonyManager) getSystemService(TELEPHONY_SERVICE);
        //系统的电话服务监听了电话状态的变化
        tm.listen(listener, PhoneStateListener.LISTEN_CALL_STATE);
    }
    private class MyPhoneListener extends PhoneStateListener {
        @Override
        public void onCallStateChanged(int state, String incomingNumber) {
            switch (state) {
            case TelephonyManager.CALL_STATE_RINGING://手机铃声正在响
                //判断 incomingNumber 是否是黑名单号码
                int mode = dao.findNumberMode(incomingNumber);
                if (mode == 0 || mode == 2) {
```

```java
                //黑名单号码
                Log.i(TAG, "挂断电话");
                //挂断电话
                endcall(incomingNumber);
            }
            break;
        case TelephonyManager.CALL_STATE_IDLE: //手机的空闲状态
            break;
        case TelephonyManager.CALL_STATE_OFFHOOK://手机接通通话的状态
            break;
        }
        super.onCallStateChanged(state, incomingNumber);
    }
}
/**
 * 取消电话状态的监听
 */
@Override
public void onDestroy() {
    super.onDestroy();
    tm.listen(listener, PhoneStateListener.LISTEN_NONE);
    listener = null;
}
/**
 * 挂断电话
 * 需要复制两个 aidl 文件
 * 添加权限<uses-permission android:name="android.permission.CALL_PHONE"/>
 * @param incomingNumber
 */
public void endcall(String incomingNumber) {
    try {
        //使用反射获取系统的 service 方法
        Method method = Class.forName("android.os.ServiceManager")
                .getMethod("getService", String.class);
        IBinder binder = (IBinder) method.invoke(null,
                new Object[] { TELEPHONY_SERVICE });
        //通过 aidl 实现方法的调用
        ITelephony telephony = ITelephony.Stub.asInterface(binder);
        telephony.endCall();//该方法是一个异步方法，它会新开启一个线程将呼入
                            的号码存入数据库中
        deleteCallLog(incomingNumber);
    } catch (Exception e) {
        e.printStackTrace();
    }
}
/**
```

```
 * 删除呼叫记录
 *
 * @param incomingNumber
 */
private void deleteCallLog(String incomingNumber) {
    //呼叫记录内容提供者对应的uri
    Uri uri = Uri.parse("content://call_log/calls");
    Cursor cursor = getContentResolver().query(uri, new String[] { "_id" },
            "number=?", new String[] { incomingNumber }, null);
    while (cursor.moveToNext()) {
        String id = cursor.getString(0);
        getContentResolver().delete(uri, "_id=?", new String[] { id });
    }
    cursor.close();
}
```

测试运行：在运行前，应加上联系人的写权限<uses-permission android:name= "android.permission.WRITE_CONTACTS" />与<uses-permission android: name="android.permission.CALL_PHONE" />，此时，再次呼入黑名单中的电话拦截号码100008，通话记录如图4-22所示。

图 4-22

由图 4-22 我们可以发现，之前的黑名单电话拦截的呼叫记录全部消失，只保留了当前的黑名单号码（即使多次呼叫）。出现这种现象的原因在于：endCall()方法是一个异步的方法，在挂断电话时，同时开启了一个新的线程用于存储当前拦截掉的号码到数据库中。在执行下面的 deleteCallLog(incomingNumber)方法时，该号码还没有被存入数据库中，导致该号码最终存活了下来。

4.5 采用内容观察者删除呼叫记录

解决上节的问题可以使用内容观察者来观察数据库中的内容是否发生改变，当发现内容变化时，应立即执行删除操作。此时，**CallFirewallService.java** 中的业务代码如下：

```java
package com.guoshisp.mobilesafe.service;
import java.lang.reflect.Method;
import android.app.Service;
import android.content.Intent;
import android.database.ContentObserver;
import android.database.Cursor;
import android.net.Uri;
import android.os.Handler;
import android.os.IBinder;
import android.provider.CallLog;
import android.telephony.PhoneStateListener;
import android.telephony.TelephonyManager;
import android.util.Log;
import com.android.internal.telephony.ITelephony;
import com.guoshisp.mobilesafe.db.dao.BlackNumberDao;
public class CallFirewallService extends Service {
    public static final String TAG = "CallFirewallService";
    private TelephonyManager tm;
    private MyPhoneListener listener;
    private BlackNumberDao dao;
    @Override
    public IBinder onBind(Intent intent) {
        return null;
    }
    /**
     * 当服务第一次被创建时调用
     */
    @Override
    public void onCreate() {
        super.onCreate();
        dao = new BlackNumberDao(this);
        //注册系统的电话状态改变的监听器
        listener = new MyPhoneListener();
        tm = (TelephonyManager) getSystemService(TELEPHONY_SERVICE);
        //系统的电话服务监听了电话状态的变化
        tm.listen(listener, PhoneStateListener.LISTEN_CALL_STATE);
    }
    private class MyPhoneListener extends PhoneStateListener {
```

```java
    @Override
    public void onCallStateChanged(int state, String incomingNumber) {
        switch (state) {
        case TelephonyManager.CALL_STATE_RINGING://手机铃声正在响
            //starttime = System.currentTimeMillis();
            //判断 incomingNumber 是否是黑名单号码
            int mode = dao.findNumberMode(incomingNumber);
            if (mode == 0 || mode == 2) {
                //黑名单号码
                Log.i(TAG, "挂断电话");
                //挂断电话
                endcall(incomingNumber);
            }
            break;
        case TelephonyManager.CALL_STATE_IDLE: //手机的空闲状态
            break;
        case TelephonyManager.CALL_STATE_OFFHOOK://手机接通通话的状态
            break;
        }
        super.onCallStateChanged(state, incomingNumber);
    }
}
/**
 * 取消电话状态的监听
 */
@Override
public void onDestroy() {
    super.onDestroy();
    tm.listen(listener, PhoneStateListener.LISTEN_NONE);
    listener = null;
}
/**
 * 挂断电话
 * 需要复制两个 aidl 文件
 * 添加权限<uses-permission android:name="android.permission.CALL_PHONE"/>
 * @param incomingNumber
 */
public void endcall(String incomingNumber) {
    try {
        //使用反射获取系统的 service 方法
        Method method = Class.forName("android.os.ServiceManager")
                .getMethod("getService", String.class);
        IBinder binder = (IBinder) method.invoke(null,
                new Object[] { TELEPHONY_SERVICE });
        //通过 aidl 实现方法的调用
        ITelephony telephony = ITelephony.Stub.asInterface(binder);
```

```java
            telephony.endCall();//该方法是一个异步方法，会新开启一个线程将呼入的
                              号码存入数据库中
                //deleteCallLog(incomingNumber);
            //注册一个内容观察者,观察uri数据的变化
            getContentResolver().registerContentObserver(
                    CallLog.Calls.CONTENT_URI, true, new MyObserver(new
                        Handler(), incomingNumber));
        } catch (Exception e) {
            e.printStackTrace();
        }
    }
    /**
     * 定义自己的内容观察者
     * 在构造方法里面传递要观察的号码
     * @author
     *
     */
    private class MyObserver extends ContentObserver {
        private String incomingNumber;
        public MyObserver(Handler handler, String incomingNumber) {
            super(handler);
            this.incomingNumber = incomingNumber;
        }
        /**
         * 数据库内容发生改变时调用的方法
         */
        @Override
        public void onChange(boolean selfChange) {
            super.onChange(selfChange);
            //立即执行删除操作
            deleteCallLog(incomingNumber);
            //停止数据的观察
            getContentResolver().unregisterContentObserver(this);
        }
    }
    /**
     * 删除呼叫记录
     *
     * @param incomingNumber
     */
    private void deleteCallLog(String incomingNumber) {
        //呼叫记录内容提供者对应的uri
        Uri uri = Uri.parse("content://call_log/calls");
        Cursor cursor = getContentResolver().query(uri, new String[] { "_id" },
                "number=?", new String[] { incomingNumber }, null);
        while (cursor.moveToNext()) {
```

```
            String id = cursor.getString(0);
            getContentResolver().delete(uri, "_id=?", new String[] { id });
        }
        cursor.close();
    }
}
```

测试运行：再次呼入 100008，通话记录如图 4-23 所示。

图 4-23

需要掌握的知识点小结

（1）对数据库的增、删、改、查操作。

（2）借助广播实现对黑名单短信的拦截。

（3）使用 TelephoneyManager 监听系统电话的状态。

（4）listview 的优化（复用 convertview，采用 viewholder）。

（5）界面上实现数据库的添加删除和修改，调用 adapter.notifyDataSetChange()实现列表中的数据及时更新。

（6）使用上、下文菜单为 ListView 中的 Item 设置长按后的界面显示。

（7）通过 aidl 方式获取系统的电话管理服务。

（8）通过内容观察者观察一个 uri 数据的改变来及时更新数据库中的数据。

第 5 章　其他模块的设计

5.1　软件管理模块设计

需求：当进入"软件管理"界面后，在界面的最上方显示出当前手机的可用内存和 Sdcard 的可用容量，在对应用进行展示时区分出用户程序和系统程序。展示应用程序的每个 Item 都包含：应用程序的图标、版本号、应用名称信息。当单击 Item 时，会以 PopupWindow 的样式弹出一个提示对话框：卸载、启动、分享该应用程序，单击这些选项时，会执行对应的动作。

实现思路：通过 PackageManager 获得手机中所有的应用程序对象；通过 StatFs 获取手机、Sdcard 下有多少块分区，以及每块分区的大小，最终通过工具类 Formatter 来计算出大小；在 ListView 中为 Item 注册单击事件，触发该单击事件时会弹出一个 PopupWindow 来提示当前程序的卸载、启动、分享的操作，这些动作是通过调用系统的意图来实现的。软件管理的 UI 设计如图 5-1 所示。

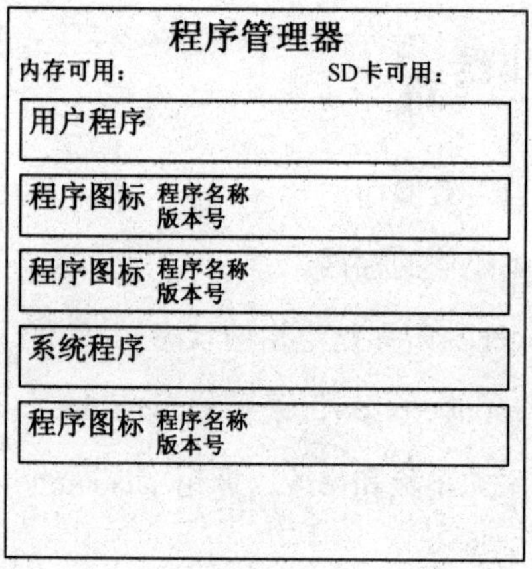

图 5-1

5.1.1　软件管理器之分类显示应用程序

首先，我们实现图 5-1 的分类展示。该界面对应的 Activity 为 AppManagerActivity，AppManagerActivity.java 的业务代码如下：

```java
package com.guoshisp.mobilesafe;
import java.io.File;
import java.util.ArrayList;
import java.util.List;
import android.app.Activity;
import android.content.pm.PackageManager;
import android.os.Bundle;
import android.os.Environment;
import android.os.Handler;
import android.os.Message;
import android.os.StatFs;
import android.text.format.Formatter;
import android.view.View;
import android.view.ViewGroup;
import android.widget.BaseAdapter;
import android.widget.ImageView;
import android.widget.LinearLayout;
import android.widget.ListView;
import android.widget.TextView;
import com.guoshisp.mobilesafe.domain.AppInfo;
import com.guoshisp.mobilesafe.engine.AppInfoProvider;
public class AppManagerActivity extends Activity{
    protected static final int LOAD_APP_FINSISH = 50;
    private static final String TAG = "AppManagerActivity";
    private TextView tv_appmanager_mem_avail;//显示手机可用内存
    private TextView tv_appmanager_sd_avail;//显示Sdcard可用内存
    private ListView lv_appmanager;//展示用户程序、系统程序
    private LinearLayout ll_loading;//ProgressBar的父控件，用于控制该控件中
                                    //的子控件的显示
    private PackageManager pm; //相当于windows系统下面的程序管理器（可以获取
                               //手机中所有的应用程序）
    private List<AppInfo> appinfos;//存放手机中所有应用程序（用户程序+系统程序）
    private List<AppInfo> userappInfos;//存放用户程序
    private List<AppInfo> systemappInfos;//存放系统程序
    //当应用程序在子线程中全部加载成功后，通知主线程显示数据
    private Handler handler = new Handler() {
        public void handleMessage(android.os.Message msg) {
            switch (msg.what) {
            case LOAD_APP_FINSISH:
                ll_loading.setVisibility(View.INVISIBLE);
                lv_appmanager.setAdapter(new AppManagerAdapter());
                break;
            }
        };
    };
    @Override
```

```java
protected void onCreate(Bundle savedInstanceState) {
    setContentView(R.layout.app_manager);
    super.onCreate(savedInstanceState);
    tv_appmanager_mem_avail = (TextView) findViewById(R.id.tv_appmanager_
                    mem_avail);
    tv_appmanager_sd_avail = (TextView) findViewById(R.id.tv_appmanager_
                    sd_avail);
    ll_loading = (LinearLayout) findViewById(R.id.ll_appmanager_loading);
    lv_appmanager = (ListView) findViewById(R.id.lv_appmanager);
    tv_appmanager_sd_avail.setText("SD卡可用" + getAvailSDSize());
    tv_appmanager_mem_avail.setText("内存可用:" + getAvailROMSize());
    pm = getPackageManager();
    //加载所有应用程序的数据
    fillData();
}
/**
 * 将手机中的应用程序全部获取出来
 */
private void fillData() {
    //加载数据时，ll_loading控件中的ProgressBar以及TextView对应的"正在
        加载数据..."显示出来
    ll_loading.setVisibility(View.VISIBLE);
    new Thread() {
        public void run() {
            AppInfoProvider provider = new AppInfoProvider(
                    AppManagerActivity.this);
            appinfos = provider.getInstalledApps();
            initAppInfo();
            //向主线程发送消息
            Message msg = Message.obtain();
            msg.what = LOAD_APP_FINSISH;
            handler.sendMessage(msg);
        };
    }.start();
}
/**
 * 初始化系统和用户appinfos的集合
 */
protected void initAppInfo() {
    systemappInfos = new ArrayList<AppInfo>();
    userappInfos = new ArrayList<AppInfo>();
    for (AppInfo appinfo : appinfos) {
        //区分出用户程序和系统程序
        if (appinfo.isUserapp()) {
            userappInfos.add(appinfo);
        } else {
```

```java
            systemappInfos.add(appinfo);
        }
    }
}
//适配器对象
private class AppManagerAdapter extends BaseAdapter {
    //获取ListView中Item的数据
    public int getCount() {
        //因为listview要多显示两个条目（用户程序和系统程序）
        return userappInfos.size() + 1 + systemappInfos.size() + 1;
    }
    //获取到Item所对应的对象
    public Object getItem(int position) {
        //当position == 0 则对应的是"用户程序"条目
        if (position == 0) {
            return position;
        } else if (position <= userappInfos.size()) {//当position <=
                userappInfos.size()则对应的是手机中所有用户程序条目
            // 要显示的用户程序的条目在集合中的位置（因为用户程序对应的Item是
                从1开始的，而集合中的角标是从0开始的）
            int newpostion = position - 1;
            return userappInfos.get(newpostion);
        } else if (position == (userappInfos.size() + 1)) {//当position
                == (userappInfos.size() + 1)则对应的是"系统程序"条目
            return position;
        } else {//所有系统应用条目
            int newpostion = position - userappInfos.size() - 2;
            return systemappInfos.get(newpostion);
        }
    }
    //获取Item所对应的id
    public long getItemId(int position) {
        return position;
    }
    //将View显示在Item上，每显示一个Item，调用一次该方法
    public View getView(int position, View convertView, ViewGroup
                parent) {
        if (position == 0) {//如果position == 0，则对应的是"用户程序"条目，
                我们创建出该条目对应的View
            TextView tv = new TextView(getApplicationContext());
            tv.setTextSize(20);
            tv.setText("用户程序 (" + userappInfos.size() + ")");
            return tv;
        } else if (position <= userappInfos.size()) {//当position <=
                userappInfos.size()则对应的是手机中所有用户程序条目
            //要显示的用户程序的条目在集合中的位置（因为用户程序对应的Item是
```

从 1 开始的，而集合中的角标是从 0 开始的）

```java
    int newpostion = position - 1;
    View view;
    ViewHolder holder;
    //复用历史缓存
    if (convertView == null || convertView instanceof TextView) {
        view = View.inflate(getApplicationContext(),
                R.layout.app_manager_item, null);
        holder = new ViewHolder();
        holder.iv_icon = (ImageView) view
                .findViewById(R.id.iv_appmanger_icon);
        holder.tv_name = (TextView) view
                .findViewById(R.id.tv_appmanager_appname);
        holder.tv_version = (TextView) view
                .findViewById(R.id.tv_appmanager_appversion);
        view.setTag(holder);
    } else {
        view = convertView;
        holder = (ViewHolder) view.getTag();
    }
    //为用户应用程序适配数据
    AppInfo appInfo = userappInfos.get(newpostion); //从用户程
                            序集合里面获取数据的条目
    holder.iv_icon.setImageDrawable(appInfo.getAppicon());
    holder.tv_name.setText(appInfo.getAppname());
    holder.tv_version.setText("版本号:" + appInfo.getVersion());
    return view;
} else if (position == (userappInfos.size() + 1)) {//如果position
         == (userappInfos.size() + 1)则对应的是"系统程序"条目
    TextView tv = new TextView(getApplicationContext());
    tv.setTextSize(20);
    tv.setText("系统程序 (" + systemappInfos.size() + ")");
    return tv;
} else {//所有系统应用的Item
    int newpostion = position - userappInfos.size() - 2;
    View view;
    ViewHolder holder;
    if (convertView == null || convertView instanceof TextView) {
        view = View.inflate(getApplicationContext(),
                R.layout.app_manager_item, null);
        holder = new ViewHolder();
        holder.iv_icon = (ImageView) view
                .findViewById(R.id.iv_appmanger_icon);
        holder.tv_name = (TextView) view
                .findViewById(R.id.tv_appmanager_appname);
        holder.tv_version = (TextView) view
```

```java
                    .findViewById(R.id.tv_appmanager_appversion);
            view.setTag(holder);
        } else {
            view = convertView;
            holder = (ViewHolder) view.getTag();
        }
        //为系统应用程序适配数据
        AppInfo appInfo = systemappInfos.get(newpostion); //从系
                            统程序集合里面获取数据的条目
        holder.iv_icon.setImageDrawable(appInfo.getAppicon());
        holder.tv_name.setText(appInfo.getAppname());
        holder.tv_version.setText("版本号:" + appInfo.getVersion());
        return view;
    }
}
/**
 * 屏蔽掉连个textview的单击事件
 *//*
@Override
public boolean isEnabled(int position) {
    if (position == 0 || position == (userappInfos.size() + 1)) {
        return false;
    }
    return super.isEnabled(position);
}*/
}
//将Item中的控件使用static修饰，被static修饰的类的字节码在JVM中只会存在一
  份。iv_icon, tv_name与tv_version在栈中也会只存在一份
private static class ViewHolder {
    ImageView iv_icon;
    TextView tv_name;
    TextView tv_version;
}
/**
 * 获取Sdcard卡可用的内存大小
 *
 * @return
 */
private String getAvailSDSize() {
    //获取Sdcard根目录所在的文件对象
    File path = Environment.getExternalStorageDirectory();
    //状态空间对象
    StatFs stat = new StatFs(path.getPath());
    //获取Sdcard卡中有多少块分区（整个Sdcard的空间被分为多块）
    long totalBlocks = stat.getBlockCount();
    //获取Sdcard卡可用的分区数量
```

```java
        long availableBlocks = stat.getAvailableBlocks();
        //获取 Sdcard 卡每一块分区可以存放的 byte 数量
        long blockSize = stat.getBlockSize();
        //计算中总的 byte
        long availSDsize = availableBlocks * blockSize;
        //借助 Formatter 来将其转换为 MB
        return Formatter.formatFileSize(this, availSDsize);
    }
    /**
     * 获取手机剩余可用的内存空间
     *
     * @return
     */
    private String getAvailROMSize() {
        File path = Environment.getDataDirectory();
        StatFs stat = new StatFs(path.getPath());
        long blockSize = stat.getBlockSize();
        long availableBlocks = stat.getAvailableBlocks();
        return Formatter.formatFileSize(this, availableBlocks * blockSize);
    }
}
```

注意:"用户程序"条目和"系统程序"条目都是 ListView 中的 Item。

代码解析:

if (convertView == null || convertView instanceof TextView)复用历史缓存的 View 对象,首先要判断该对象是否为 null。这里,如果不判断缓存 View 对象的类型,当滑动 ListView 时,程序会抛出 NullPointeException。原因在于:position==0 代表的是"用户程序"对应的条目,这是我们创建的一个 TextView,创建并赋值之后直接执行 return tv,而没有执行 view.setTag(holder),也就没有将其放入到缓存中。即使将其存入到缓存中,也不能够复用该 View,因为该 View 仅仅是用于显示一个文本而存在的。

该 Activity 对应的布局文件存放在 layout 目录下,app_manager.xml 文件内容如下:

```xml
<?xml version="1.0" encoding="utf-8"?>
<LinearLayout xmlns:android="http://schemas.android.com/apk/res/android"
    android:layout_width="fill_parent"
    android:layout_height="fill_parent"
    android:orientation="vertical" >
    <TextView
        android:layout_width="wrap_content"
        android:layout_height="wrap_content"
        android:layout_gravity="center_horizontal"
        android:text="程序管理器"
        android:textColor="#66ff00"
        android:textSize="28sp" />
```

```xml
<View
    android:layout_width="fill_parent"
    android:layout_height="1dip"
    android:layout_marginTop="5dip"
    android:background="@drawable/devide_line" />
<RelativeLayout
    android:layout_width="fill_parent"
    android:layout_height="wrap_content"
    android:layout_marginTop="5dip" >
    <TextView
        android:id="@+id/tv_appmanager_mem_avail"
        android:layout_width="wrap_content"
        android:layout_height="wrap_content"
        android:text="内存可用"
        android:textSize="16sp" />
    <TextView
        android:id="@+id/tv_appmanager_sd_avail"
        android:layout_width="wrap_content"
        android:layout_height="wrap_content"
        android:layout_alignParentRight="true"
        android:text="SD 卡可用"
        android:textSize="16sp" />
</RelativeLayout>
<RelativeLayout
    android:layout_width="fill_parent"
    android:layout_height="fill_parent" >
    <LinearLayout
        android:id="@+id/ll_appmanager_loading"
        android:layout_width="fill_parent"
        android:layout_height="fill_parent"
        android:gravity="center"
        android:orientation="vertical"
        android:visibility="invisible" >
        <ProgressBar
            android:layout_width="wrap_content"
            android:layout_height="wrap_content" />

        <TextView
            android:layout_width="wrap_content"
            android:layout_height="wrap_content"
            android:text="正在加载程序信息..." />
    </LinearLayout>
    <ListView
        android:id="@+id/lv_appmanager"
        android:layout_width="fill_parent"
        android:layout_height="fill_parent" >
```

```xml
        </ListView>
    </RelativeLayout>
</LinearLayout>
```

每个 Item 对应的布局文件也存放在 layout 目录下，app_manager_item.xml 文件内容如下：

```xml
<?xml version="1.0" encoding="utf-8"?>
<RelativeLayout xmlns:android="http://schemas.android.com/apk/res/android"
    android:layout_width="match_parent"
    android:layout_height="wrap_content" >
    <ImageView
        android:id="@+id/iv_appmanger_icon"
        android:layout_width="60dip"
        android:layout_height="60dip"
        android:src="@drawable/ic_launcher" />
    <TextView
        android:singleLine="true"
        android:ellipsize="end"
        android:id="@+id/tv_appmanager_appname"
        android:layout_width="wrap_content"
        android:layout_height="wrap_content"
        android:layout_toRightOf="@id/iv_appmanger_icon"
        android:text="程序名称"
        android:textSize="24sp" />
    <TextView
        android:id="@+id/tv_appmanager_appversion"
        android:layout_width="wrap_content"
        android:layout_height="wrap_content"
        android:layout_below="@id/tv_appmanager_appname"
        android:layout_marginTop="8dip"
        android:layout_toRightOf="@id/iv_appmanger_icon"
        android:text="版本号:"
        android:textSize="14sp" />
</RelativeLayout>
```

其中，android:singleLine="true"表示的是显示一行文本，android:ellipsize= "end"表示多出的文本用"…"替代，替代文本的尾部。

在 AppManagerActivity 中获取应用程序的数据是通过 AppInfoProvider 获取的，该对象位于包"AppInfoProvider"下，AppInfoProvider.java 代码如下：

```java
package com.guoshisp.mobilesafe.engine;
import java.util.ArrayList;
import java.util.List;
import android.content.Context;
import android.content.pm.ApplicationInfo;
import android.content.pm.PackageInfo;
import android.content.pm.PackageManager;
```

```java
import com.guoshisp.mobilesafe.domain.AppInfo;
public class AppInfoProvider {
    private PackageManager pm;
    public AppInfoProvider(Context context) {
        pm = context.getPackageManager();
    }
    /**
     * 获取所有的安装的程序信息
     * @return
     */
    public List<AppInfo> getInstalledApps(){//返回所有的安装的程序列表信息
              其中，参数 PackageManager.GET_UNINSTALLED_PACKAGES 表示包
              括那些被卸载但是没有清除数据的应用
        List<PackageInfo> packageinfos = pm.getInstalledPackages
                    (PackageManager.GET_UNINSTALLED_PACKAGES);
        List<AppInfo> appinfos = new ArrayList<AppInfo>();
        for(PackageInfo info : packageinfos){
            AppInfo appinfo = new AppInfo();
            //应用程序的包名
            appinfo.setPackname(info.packageName);
            //应用程序的版本号
            appinfo.setVersion(info.versionName);
            //应用程序的图标 info.applicationInfo.loadIcon(pm);
            appinfo.setAppicon(info.applicationInfo.loadIcon(pm));
            //应用程序的名称 info.applicationInfo.loadLabel(pm);
appinfo.setAppname(info.applicationInfo.loadLabel(pm).toString());
            //过滤出用户应用程序的名称
            appinfo.setUserapp(filterApp(info.applicationInfo));
            appinfos.add(appinfo);
            appinfo = null;
        }
        return appinfos;
    }
    /**
     * 第三方应用程序的过滤器
     * @param info
     * @return true 三方应用
     *  false 系统应用
     */
    public boolean filterApp(ApplicationInfo info) {
    //当前应用程序的标记与系统应用程序的标记
        if ((info.flags & ApplicationInfo.FLAG_UPDATED_SYSTEM_APP) != 0) {
            return true;
        } else if ((info.flags & ApplicationInfo.FLAG_SYSTEM) == 0) {
            return true;
        }
```

```
            return false;
    }
}
```

用于封装应用信息的是 AppInfo 对象，AppInfo.java 代码如下：

```java
package com.guoshisp.mobilesafe.domain;
import android.graphics.drawable.Drawable;
public class AppInfo {
    //应用包名
    private String packname;
    //应用版本号
    private String version;
    //应用名称
    private String appname;
    //应用图标
    private Drawable appicon;
    //该应用是否属于用户程序
    private boolean userapp;
    public boolean isUserapp() {
        return userapp;
    }
    public void setUserapp(boolean userapp) {
        this.userapp = userapp;
    }
    public String getPackname() {
        return packname;
    }
    public void setPackname(String packname) {
        this.packname = packname;
    }
    public String getVersion() {
        return version;
    }
    public void setVersion(String version) {
        this.version = version;
    }
    public String getAppname() {
        return appname;
    }
    public void setAppname(String appname) {
        this.appname = appname;
    }
    public Drawable getAppicon() {
        return appicon;
    }
    public void setAppicon(Drawable appicon) {
```

```
        this.appicon = appicon;
    }
}
```

测试运行：进入程序主界面后，单击"软件管理"后，进入"程序管理"界面，界面如图 5-2 所示。

图 5-2

5.1.2 使用 PopupWindow 显示程序的启动、分享、卸载

需求：当单击 Item 时，在该 Item 对应位置弹出一个 PopupWindow 窗体，在该窗体中显示"启动、分享、卸载"。代码见在线资源包中代码文本部分 5.1.2.doc。

代码解析：

（1）dismissPopupWindow()

该方法执行关闭 PopupWindow 窗口的操作，如果不执行该方法，当单击其他 Item 时，将会出现多个 PopupWindow 存在的情况。

（2）popupWindow = new PopupWindow(contentView,DensityUtil.dip2px (getApplicationContext(), 200), bottom - top+ DensityUtil.dip2px (getApplicationContext(), 20))

创建 PopupWindow 窗口，参数二、参数三指定 PopupWindow 窗口的大小，长为 200px，高度为 Item 的高度（如果不指定大小，PopupWindow 窗口是不会被显示到界面上的）。这里使用了一个工具类 DensityUtil，将 dp 转成 px，这样做的好处是解决了 PopupWindow 在不同分辨率的手机上运行时出现的屏幕适配问题。该工具类位于包"com.guoshisp.mobilesafe.utils"下，DensityUtil.java 工具代码如下：

```
package com.guoshisp.mobilesafe.utils;
```

```java
import android.content.Context;
public class DensityUtil {
    /**
     * 根据手机的分辨率从 dp 的单位转为 px(像素)
     */
    public static int dip2px(Context context, float dpValue) {
        final float scale = context.getResources().getDisplayMetrics().
                             density;
        return (int) (dpValue * scale + 0.5f);
    }
    /**
     * 根据手机的分辨率从 px(像素) 的单位转为 dp
     */
    public static int px2dip(Context context, float pxValue) {
        final float scale = context.getResources().getDisplayMetrics().
                             density;
        return (int) (pxValue / scale + 0.5f);
    }
}
```

（3）popupWindow.setBackgroundDrawable(new ColorDrawable(
　　　　　　　　Color.TRANSPARENT))

用于为 PopupWindow 设置背景资源。如果不为其设置背景资源，PopupWindow 的焦点获取和动画的播放都会出现问题。

（4）popupWindow.showAtLocation(view, Gravity.TOP | Gravity.LEFT,
　　　　　　　　location[0] + 20, location[1])

设置 PopupWindow 窗口显示的位置。参数一：当前的 PopupWindow 要挂载到哪个 view 对象上；参数二：设置 PopupWindow 显示的重心位置；参数三、参数四：PopupWindow 在 X 轴、Y 轴的偏移量，偏移的参考点是当前 Activity 所在窗体的（0, 0）位置。这里是根据当前 Item 在屏幕中位置而确定的，因为 PopupWindow 需要显示到对应的应用程序条目上。

（5）View contentView = View.inflate(getApplicationContext(),
　　　　　　　　R.layout.popup_item, null)

获取 PopupWindow 中用于显示内容的 contentView 对象，popup_item.xml 文件内容如下：

```xml
<?xml version="1.0" encoding="utf-8"?>
<LinearLayout xmlns:android="http://schemas.android.com/apk/res/android"
    android:layout_width="wrap_content"
    android:layout_height="wrap_content"
    android:background="@drawable/local_popup_bg"
    android:id="@+id/ll_popup_container"
    android:orientation="horizontal" >
    <LinearLayout
        android:id="@+id/ll_popup_uninstall"
```

```xml
        android:layout_width="wrap_content"
        android:layout_height="wrap_content"
        android:gravity="center_horizontal"
        android:orientation="vertical" >
        <ImageView
            android:layout_width="wrap_content"
            android:layout_height="wrap_content"
            android:src="@drawable/img1" />
        <TextView
            android:layout_width="wrap_content"
            android:layout_height="wrap_content"
            android:text="卸载" />
</LinearLayout>
<LinearLayout
    android:id="@+id/ll_popup_start"
    android:layout_width="wrap_content"
    android:layout_height="wrap_content"
    android:layout_marginLeft="5dip"
    android:gravity="center_horizontal"
    android:orientation="vertical" >
    <ImageView
        android:layout_width="wrap_content"
        android:layout_height="wrap_content"
        android:src="@drawable/img2" />
    <TextView
        android:layout_width="wrap_content"
        android:layout_height="wrap_content"
        android:text="启动" />
</LinearLayout>
<LinearLayout
    android:id="@+id/ll_popup_share"
    android:layout_width="wrap_content"
    android:layout_height="wrap_content"
    android:layout_marginLeft="5dip"
    android:gravity="center_horizontal"
    android:orientation="vertical" >
    <ImageView
        android:layout_width="wrap_content"
        android:layout_height="wrap_content"
        android:src="@drawable/img3" />
    <TextView
        android:layout_width="wrap_content"
        android:layout_height="wrap_content"
        android:text="分享" />
```

```
        </LinearLayout>
    </LinearLayout>
```

该文件的预览图如图 5-3 所示。

测试运行：单击任意一个 Item 的界面效果图如图 5-4 所示，分别单击 PopupWindow 中的"卸载、启动、分享"后，Log 日志信息如图 5-5 所示。

图 5-3

图 5-4

图 5-5

5.1.3　实现程序的卸载、启动、分享功能

实现程序的卸载、启动、分享，可以通过意图来实现。当卸载掉一个用户程序（禁止卸载系统应用程序）后，应当更新当前界面中的内存、Sdcard、显示条目的信息。

AppManagerActivity.java 的最终业务代码见在线资源包中代码文本部分 5.1.3.doc。

分享对应的意图过滤器为：

```
<intent-filter>
    <action android:name="android.intent.action.SEND" />
    <category android:name="android.intent.category.DEFAULT" />
```

```
            <data android:mimeType="text/plain" />
    </intent-filter>
```

卸载对应的意图过滤器为：

```
<intent-filter>
        <action android:name="android.intent.action.VIEW" />
        <action android:name="android.intent.action.DELETE" /> <category
                        android:name="android.intent.category.
                        DEFAULT"/> <data android:scheme="package" />
</intent-filter>*/
```

测试运行：单击一个用户程序对应的 Item，卸载后，展示的界面如图 5-6 所示；单击分享后，展示的界面如图 5-7 所示（在真实手机上会展示出具有分享功能的应用供我们选择）；单击计算器对应的 Item，启动后，对应的界面如图 5-8 所示。

图 5-6　　　　　　　　　　图 5-7　　　　　　　　　　图 5-8

需要掌握的知识点小结

（1）使用 PackageManager 获取系统中的所有应用程序。

（2）通过 StatFs 与 Formatter 来获取手机内存、Sdcard 的的可用内存。

（3）自定义的 Adapter 实现复杂 View 的展示。

（4）使用 PopupWindow 需要注意：要为其设置大小和背景资源。

（5）通过系统的意图来实现软件的卸载、启动、分享功能。

（6）ListView 的优化。

（7）ListView 的单击、滑动事件的监听。

5.2 进程管理器的设计

需求：进入到进程管理器后，我们会将手机中的进程分为两类——用户进程与系统进程。当用户单击"用户进程"按钮时，将所有的用户进程显示在列表中；当用户单击"系统进程"按钮时，将所有的系统进程显示在列表中。在显示进程信息的 Item 中，会显示每个进程所对应的图标、包名、所占内存，在 Item 的右端有一个 Checkbox 的勾选框。屏幕的底部存在两个按钮——全选、一键清理。当单击"全选"按钮后，当前列表中的 Checkbox 都被勾选上。在单击"一键清理"按钮后，会将当前列表中已勾选的 Checkbox 对应的进程全部杀死，同时使用自定义的 Toast 来提示杀死了多少个进程、释放了多少内存空间，并将杀死的进程从当前的列表中移除。

实现思路：使用 ActivityManager 获取到当前正在运行的所有进程，然后迭代出每个进程对应的包名，获取到包名后，再通过 PackageManager 来获取到该包名所对应的应用的对象，拿到应用对象后就可以得到该应用中的所有属性信息。

显示用户进程时对应的 UI 界面如图 5-9 所示，显示系统进程时对应的 UI 界面如图 5-10 所示。

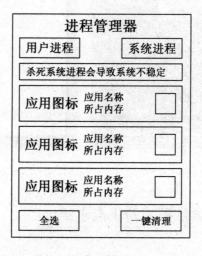

图 5-9　　　　　　　　　　　　　　图 5-10

实现说明：在讲解"程序锁"应用时，已经向读者展示过如何获取手机中的程序信息了，所以，下面将这个实现过程省略掉，将业务代码一次性展示出。

5.2.1 进程管理器的实现

进程管理器所对应的 Activity 为 TaskManagerActivity，TaskManagerActivity.java 业务代码见在线资源包中代码文本部分 5.2.1.doc。

该界面对应的布局文件 task_manager.xml 如下：

```xml
<?xml version="1.0" encoding="utf-8"?>
<LinearLayout xmlns:android="http://schemas.android.com/apk/res/android"
    android:layout_width="fill_parent"
    android:layout_height="fill_parent"
    android:orientation="vertical" >
    <TextView
        style="@style/text_title_style"
        android:gravity="center"
        android:text="进程管理器" />
    <LinearLayout
        android:layout_width="fill_parent"
        android:layout_height="wrap_content"
        android:orientation="horizontal" >
        <Button
            android:id="@+id/bt_user"
            android:layout_width="0dip"
            android:layout_height="wrap_content"
            android:layout_weight="1"
            android:background="@drawable/bg_normal"
            android:gravity="center_horizontal"
            android:text="用户进程"
            android:textColor="#ffffff"
            android:textSize="18sp" />
        <Button
            android:id="@+id/bt_system"
            android:layout_width="0dip"
            android:layout_height="wrap_content"
            android:layout_weight="1"
            android:background="@drawable/bg_normal"
            android:gravity="center_horizontal"
            android:text="系统进程"
            android:textColor="#ffffff"
            android:textSize="18sp" />
    </LinearLayout>
    <RelativeLayout
        android:layout_width="fill_parent"
        android:layout_height="fill_parent" >
        <ListView
            android:id="@+id/lv_usertask"
            android:layout_width="fill_parent"
            android:layout_height="fill_parent"
            android:layout_marginBottom="30dip" >
        </ListView>
        <ListView
            android:visibility="gone"
```

```xml
            android:id="@+id/lv_systemtask"
            android:layout_width="fill_parent"
            android:layout_height="fill_parent"
            android:layout_marginBottom="30dip" >
        </ListView>
        <LinearLayout
            android:layout_width="fill_parent"
            android:layout_height="wrap_content"
            android:layout_alignParentBottom="true"
            android:orientation="horizontal" >
            <Button
                android:layout_width="0dip"
                android:onClick="selectAll"
                android:layout_height="wrap_content"
                android:layout_weight="1"
                android:gravity="center_horizontal"
                android:text="全选"
                android:textColor="#ffffff"
                android:textSize="18sp" />
            <Button
                android:onClick="oneKeyClear"
                android:layout_width="0dip"
                android:layout_height="wrap_content"
                android:layout_weight="1"
                android:gravity="center_horizontal"
                android:text="一键清理"
                android:textColor="#ffffff"
                android:textSize="18sp" />
        </LinearLayout>
    </RelativeLayout>
</LinearLayout>
```

Item 对应的 View 布局文件 task_manager_item.xml 如下:

```xml
<?xml version="1.0" encoding="utf-8"?>
<RelativeLayout xmlns:android="http://schemas.android.com/apk/res/android"
    android:layout_width="match_parent"
    android:layout_height="wrap_content" >
    <ImageView
        android:id="@+id/iv_taskmanger_icon"
        android:layout_width="60dip"
        android:layout_height="60dip"
        android:src="@drawable/ic_launcher" />
    <TextView
        android:singleLine="true"
        android:ellipsize="end"
        android:id="@+id/tv_taskmanager_appname"
```

```xml
        android:layout_width="wrap_content"
        android:layout_height="wrap_content"
        android:layout_toRightOf="@id/iv_taskmanger_icon"
        android:text="程序名称"
        android:textSize="24sp" />
    <TextView
        android:id="@+id/tv_taskmanager_mem"
        android:layout_width="wrap_content"
        android:layout_height="wrap_content"
        android:layout_below="@id/tv_taskmanager_appname"
        android:layout_marginTop="8dip"
        android:layout_toRightOf="@id/iv_taskmanger_icon"
        android:text="占用内存:"
        android:textSize="14sp" />
    <CheckBox
        android:focusable="false"
        android:clickable="false"
        android:layout_width="wrap_content"
        android:layout_height="wrap_content"
        android:layout_alignParentRight="true"
        android:id="@+id/cb_taskmanager"
        />
</RelativeLayout>
```

该布局文件中需要注意的一点是：</Checkbox>控件中设置有两个属性，android:focusable="false"和 android:clickable="false"。设置这两个属性的原因是：Checkbox 本身有"占据"焦点的优势，如果不设置这两个属性，在单击 Checkbox 以外的 Item 时，不会响应到 Checkbox 的单击事件，因为 Checkbox 将焦点抢走了。设置这两个属性可以屏蔽掉 Checkbox 的焦点，单击 Item 中的任意位置都可以响应到 Checkbox 的单击事件。

代码解析：

（1）if (info.isChecked()) {
　　　　info.setChecked(false);
　　　　cb.setChecked(false);
　　} else {
　　　　info.setChecked(true);
　　　　cb.setChecked(true);
　　}

这是在单击 ListView 中的 Item 时执行的代码。该代码是设置每个 info 的 Checkbox 的勾选状态。如果不在这里手动设置，且将第一个 Item 的 Checkbox 勾选上，当向下滑动 ListView 将第一个 Item 滚出屏幕时，发现最底部的 Item 的 Checkbox 也被勾选上了。出现这种现象的原因是：因为对 ListView 进行了优化——复用了缓存的 View 对象（Item），当第一个 Item 被移除屏幕时，屏幕底部的 Item 需要复用这个 Item 对应的 View 对象。所以，最底部的 Item 的 Checkbox

也会处于勾选状态。然后在 getView(int position,View convertView, ViewGroup parent)方法中执行 holder.cb.setChecked(info.isChecked())代码，就可以实现手动控制 Checkbox 的状态，而不用让系统记住 Checkbox 的状态。

（2）if (view instanceof TextView) {
　　　　return;
　　}

这是在 lv2 中的单击事件中执行的第一行代码。lv2 的第一个 Item 是我们添加的 TextView，该 TextView 用于提示用户不要轻易杀死系统进程。由于在执行 Item 的单击事件时，我们处理了 Checkbox 的状态，而 lv2 中的第一个 Item 中只存在一个 TextView，此时设置 Checkbox 的状态就会出现 NullPointeException，所以，应该屏蔽掉该 Item 的单击事件。

（3）List<ProcessInfo> killedProcessInfo = new ArrayList<ProcessInfo>()

这是在执行"一键清理"的单击事件中执行的代码。我们的需求是：在执行完"一键清理"的操作后，应当将被清理的进程从列表中移除，而这就需要通知适配器重新适配数据。适配器适配数据时是从对应的集合中（用户进程集合或系统进程集合）中获取数据的，如果要保持数据同步，集合中的数据就应当将杀死的进程数据移除，但在遍历该集合时是不能够执行移除的操作的，这会导致并发访问的异常，所以，我们重新定义了一个新的集合来缓存数据信息。

（4）lv2.addHeaderView(tvheader)

将 TextView 添加到 lv2 中，作为 lv2 的第一个 Item。需要注意的一点是：执行添加的操作一定要放在执行数据适配的前面——l lv2.setAdapter(systemadapter)。

（5）am.killBackgroundProcesses(info.getPackname())

杀死一个进程的操作，执行该操作需要在清单文件中配置权限信息：<uses-permission android:name="android.permission.KILL_BACKGROUND_PROCESSES"/>。

（6）provider = new ProcessInfoProvider(this)

获取到 provider 对象的实例，通过该对象，可以获取到当前所有正在运行的应用程序的信息。该对象存放在包"com.guoshisp.mobilesafe.engine"下，ProcessInfoProvider.java 代码如下：

```
package com.guoshisp.mobilesafe.engine;
import java.util.ArrayList;
import java.util.List;
import android.app.ActivityManager;
import android.app.ActivityManager.RunningAppProcessInfo;
import android.content.Context;
import android.content.pm.ApplicationInfo;
import android.content.pm.PackageManager;
import com.guoshisp.mobilesafe.R;
import com.guoshisp.mobilesafe.domain.ProcessInfo;
```

```java
public class ProcessInfoProvider {
    private Context context;
    public ProcessInfoProvider(Context context) {
        this.context = context;
    }
    /**
     * 返回所有的正在运行的程序信息
     * @return
     */
    public List<ProcessInfo> getProcessInfos() {
        //am 可以动态地获取应用的进程信息，相当于 PC 上的进程管理器
        ActivityManager am = (ActivityManager) context
                .getSystemService(Context.ACTIVITY_SERVICE);
        //pm 可以静态地获取到手机中的所有应用程序信息，相当于 PC 上的程序管理器
        PackageManager pm = context.getPackageManager();
        //返回所有正在运行的进程
        List<RunningAppProcessInfo> runingappsInfos = am
                .getRunningAppProcesses();
        //用于存放进程信息
        List<ProcessInfo> processInfos = new ArrayList<ProcessInfo>();
        //遍历出每个进程，并将每个进程的信息封装在 ProcessInfo 对象中，最后将所有的
          进程存放在 List<ProcessInfo>中返回
        for (RunningAppProcessInfo info : runingappsInfos) {
            //用于封装进程信息
            ProcessInfo processInfo = new ProcessInfo();
            //获取进程的 pid（进程的标记）
            int pid = info.pid;
            //将进程的 pid、processName、memsize 封装到 ProcessInfo 对象中
            processInfo.setPid(pid);
            String packname = info.processName;
            processInfo.setPackname(packname);
            //获取到该进程对应的应用程序所占用的内存空间
            long memsize = am.getProcessMemoryInfo(new int[] { pid })[0]
                    .getTotalPrivateDirty() * 1024;
            processInfo.setMemsize(memsize);
            try {
                //通过进程的 packname 来获取到该进程对应的应用程序对象（获取到应用
                  程序的对象后，就可以通过该对象获取应用程序信息）
                ApplicationInfo applicationInfo = pm.getApplicationInfo
                                    (packname, 0);
                //判断该应用程序是否是第三方应用程序，便于以后分类
                if(filterApp(applicationInfo)){
                    processInfo.setUserprocess(true);
                }else{
                    processInfo.setUserprocess(false);
                }
                //分别获取到应用程序的图标和名称，并将其封装到 ProcessInfo 对象中
```

```java
                    processInfo.setIcon(applicationInfo.loadIcon(pm));
                    processInfo.setAppname(applicationInfo.loadLabel(pm).
                            toString());
        } catch (Exception e) {
//这里会抛出一个包名未找到异常,我们将其设置为系统进程,应用图标为默认的系统图标
            e.printStackTrace();
            processInfo.setUserprocess(false);
processInfo.setIcon(context.getResources().getDrawable(R.drawable.ic_la
            uncher));
            processInfo.setAppname(packname);
        }
        processInfos.add(processInfo);
        processInfo = null;
    }
    return processInfos;
}
/**
 * 三方应用的过滤器
 *
 * @param info
 * @return true 三方应用 false 系统应用
 */
public boolean filterApp(ApplicationInfo info) {
    if ((info.flags & ApplicationInfo.FLAG_UPDATED_SYSTEM_APP) != 0) {
        return true;
    } else if ((info.flags & ApplicationInfo.FLAG_SYSTEM) == 0) {
        return true;
    }
    return false;
}
}
```

代码解析:

(1) ActivityManager am = (ActivityManager) context
         .getSystemService(Context.ACTIVITY_SERVICE)

相当于 PC 上的进程管理器,可以动态地获取应用程序信息。这里是用于获取当前正在运行的程序信息。

(2) PackageManager pm = context.getPackageManager()

相当于 PC 上的程序管理器,可以获取到每个应用程序的详细信息(只能静态获取)。通过 ActivityManager 获取到进程的包名信息,PackageManager 可以获取该包名对应的应用程序的对象。得到该对象后,可以获取该应用程序中对应的属性信息。

(3) catch (Exception e)

这里会捕获一个 NoFoundPackageNameException,出现该异常的原因是:在 Android 的系

统应用程序中，有些不是使用 Java 实现的，而是使用 C 语言实现的，使用 C 语言实现的应用程序就不会被找到该应用程序所对应的包名。

（4）ProcessInfo processInfo = new ProcessInfo()

该对象是用于存放每个进程对应的应用程序的信息。该对象存放在包"com.guoshisp.mobilesafe.domain"下，ProcessInfo.java 代码如下：

```java
package com.guoshisp.mobilesafe.domain;
import android.graphics.drawable.Drawable;
public class ProcessInfo {
    //应用程序包名
    private String packname;
    //应用程序图标
    private Drawable icon;
    //应用程序所占用的内存空间，单位是byte
    private long memsize;
    //是否属于用户进程
    private boolean userprocess;
    //进程的pid（进程的标记）
    private int pid;
    //应用程序名称
    private String appname;
    //应用程序在Item中是否处于被选中状态（默认下没有被选中）
    private boolean checked;
    public boolean isChecked() {
        return checked;
    }
    public void setChecked(boolean checked) {
        this.checked = checked;
    }
    public String getAppname() {
        return appname;
    }
    public void setAppname(String appname) {
        this.appname = appname;
    }
    public int getPid() {
        return pid;
    }
    public void setPid(int pid) {
        this.pid = pid;
    }
    public String getPackname() {
        return packname;
    }
    public void setPackname(String packname) {
```

```java
        this.packname = packname;
    }
    public Drawable getIcon() {
        return icon;
    }
    public void setIcon(Drawable icon) {
        this.icon = icon;
    }
    public long getMemsize() {
        return memsize;
    }
    public void setMemsize(long memsize) {
        this.memsize = memsize;
    }
    public boolean isUserprocess() {
        return userprocess;
    }
    public void setUserprocess(boolean userprocess) {
        this.userprocess = userprocess;
    }
}
```

最后,在程序的主界面中还需要为"进程管理"条目设置单击事件,此时,MainActivity.java 的业务代码如下:

```java
package com.guoshisp.mobilesafe;
import android.app.Activity;
import android.content.Intent;
import android.os.Bundle;
import android.view.View;
import android.widget.AdapterView;
import android.widget.AdapterView.OnItemClickListener;
import android.widget.GridView;
import com.guoshisp.mobilesafe.adapter.MainAdapter;
public class MainActivity extends Activity {
    //显示主界面中九大模块的GridView
    private GridView gv_main;
    @Override
    protected void onCreate(Bundle savedInstanceState) {
        super.onCreate(savedInstanceState);
        setContentView(R.layout.main);
        gv_main = (GridView) findViewById(R.id.gv_main);
        //为gv_main对象设置一个适配器,该适配器的作用是为每个item填充对应的数据
        gv_main.setAdapter(new MainAdapter(this));
        //为GridView对象中的item设置单击时的监听事件
        gv_main.setOnItemClickListener(new OnItemClickListener() {
            //参数一:item的父控件,也就是GridView;参数二:当前单击的item;参
```

数三：当前单击的item在GridView中的位置
//参数四：id的值为单击了GridView的哪一项对应的数值，单击了GridView
　　第9项，那id就等于8
```java
public void onItemClick(AdapterView<?> parent, View view,
        int position, long id) {
    switch (position) {
    case 0: //手机防盗
        //跳转到"手机防盗"对应的Activity界面
        Intent lostprotectedIntent = new Intent(MainActivity.
                    this,LostProtectedActivity.class);
        startActivity(lostprotectedIntent);
        break;
    case 1: //通信卫士
        Intent callSmsIntent = new Intent(MainActivity.this,
                    CallSmsSafeActivity.class);
        startActivity(callSmsIntent);
        break;
    case 2: //程序管理
        Intent appManagerIntent = new Intent(MainActivity.
                        this,AppManagerActivity.class);
        startActivity(appManagerIntent);
    case 3: //进程管理
        Intent taskManagerIntent = new Intent(MainActivity.
                        this,TaskManagerActivity.class);
        startActivity(taskManagerIntent);
        break;
    case 7://高级工具
        Intent atoolsIntent = new Intent(MainActivity.this,
                    AtoolsActivity.class);
        startActivity(atoolsIntent);
        break;
    case 8://设置中心
        //跳转到"设置中心"对应的Activity界面
        Intent settingIntent = new Intent(MainActivity.this,
                    SettingCenterActivity.class);
        startActivity(settingIntent);
        break;
    }
    }
});
}
}
```

测试运行：配置好权限及组件信息后，进入主界面单击"进程管理"条目后进入进程管理界面，如图5-11所示。单击"系统进程"按钮后的界面如图5-12所示。单击系统进程中的"全选"按钮，并单击"一键清理"按钮后的界面如图5-13所示。

图 5-11　　　　　　　　图 5-12　　　　　　　　图 5-13

5.2.2　使用自定义吐司显示清理结果

在图 5-13 中，显示的系统默认的吐司界面不够美观，考虑通过自定义吐司实现结果的显示。自定义的吐司对应的 View 存放在包 "com.guoshisp.mobilesafe.view" 下，MyToast.java 代码如下：

```java
package com.guoshisp.mobilesafe.view;
import android.content.Context;
import android.view.Gravity;
import android.view.View;
import android.widget.TextView;
import android.widget.Toast;
import com.guoshisp.mobilesafe.R;
public class MyToast {
    /**
     * 显示自定义的吐司
     * @param text 显示的内容
     */
    public static void showToast(Context context, String text) {
        Toast toast = new Toast(context);
        View view = View.inflate(context, R.layout.mytoast, null);
        TextView tv = (TextView) view.findViewById(R.id.tv_toast);
        //设置显示内容
        tv.setText(text);
        toast.setView(view);
        //设置 Toast 显示的时长。0 表示短，1 表示长
        toast.setDuration(1);
        //设置 Toast 显示在窗体中的位置（这里是显示在窗体顶部的中央）
        toast.setGravity(Gravity.TOP, 0, 0);
        //将 Toast 显示出来
```

```
        toast.show();
    }
}
```

吐司中用于显示内容的布局文件 mytoast.xml 文件如下：

```
<?xml version="1.0" encoding="utf-8"?>
<LinearLayout xmlns:android="http://schemas.android.com/apk/res/android"
    android:layout_width="match_parent"
    android:layout_height="wrap_content"
    android:gravity="center_vertical"
    android:orientation="horizontal"
    android:background="@drawable/call_locate_orange"
    >
    <ImageView
        android:layout_width="wrap_content"
        android:layout_height="wrap_content"
        android:src="@drawable/notification" />
    <TextView
        android:id="@+id/tv_toast"
        android:layout_width="wrap_content"
        android:layout_height="wrap_content"
        android:textSize="14sp"
        android:text="我是土司" />
</LinearLayout>
```

最后，只需要在 TaskManagerActivity 中将系统定义的吐司改为自定义的吐司即可（注意导入包），代码如下：

```
MyToast.showToast(this,"杀死了" + count + "个进程,释放了"
+ Formatter.formatFileSize(this, memsize) + "内存");
```

此时，再次执行系统进程中的全选→一键清理操作时的界面如图 5-14 所示。

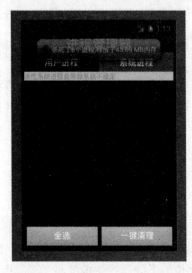

图 5-14

细心的读者可能已经发现了，系统中的大部分进程是杀不掉的（我们只是将被杀过的进程从界面中移除），市场上的主流系统管理软件也同样不能够杀死大部分系统进程，实现原理几乎一样。

需要掌握的知识点小结

（1）使用 ActivityManager 获取正在运行的程序信息（pid、占用内存大小等）。

（2）使用 PackageManager 获取对应的进程的应用程序信息。

（3）ListView 的优化。如何向 ListView 中添加一个 Item，以及添加时的注意事项。

（4）自定义吐司。

5.3 流量管理模块的设计

需求：使用 ListView 列表展示出手机中所有具有 Internet 权限的应用程序，每个 Item 展示出该应用对应的图标、应用名称、上传流量、下载流量、总流量信息。显示统计列表时，需要将 ListView 从底部拉上来才可以看见具体应用的流量信息（类似于一个抽屉），在加载应用程序的流量信息时，显示加载的 ProgressBar 是自定义的。

流量管理对应的 UI 如图 5-15 所示。

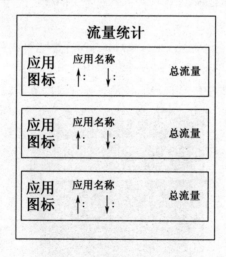

图 5-15

5.3.1 流量统计的原理

自从 Android2.2 开始，Google 为其添加了流量统计的 API-TrafficStats，通过该 API，我们可以获取到手机中应用程序的上传与下载所产生的流量信息。其实 TrafficStats 对象本身就是读取 Linux 提供的文件对象系统类型的文本进行解析得到数据。在 android.net.TrafficStats 类中，提供了多种静态方法，可以直接调用，返回类型均为 long 类型（对应的流量单位为 KB），如果返回值等于−1，代表 UNSUPPORTED，当前设备不支持统计（模拟器是不支持流量统计的，在此，我们使用真机进行测试）。API 及其说明如下：

- getMobileRxBytes()：获取通过 Mobile 连接收到的字节总数，不包含 WiFi；
- getMobileRxPackets()：/获取 Mobile 连接收到的数据包总数；
- getMobileTxBytes()：获取 Mobile 发送的总字节数；
- getMobileTxPackets()：获取 Mobile 发送的总数据包数；
- getTotalRxBytes()：获取总的接受字节数，包含 Mobile 和 WiFi 等；
- getTotalRxPackets()：总的接受数据包数，包含 Mobile 和 WiFi 等；
- getTotalTxBytes()：总的发送字节数，包含 Mobile 和 WiFi 等；
- getTotalTxPackets()：发送的总数据包数，包含 Mobile 和 WiFi 等；
- getUidRxBytes(int uid)：获取某个网络 UID 的接收的字节数；
- getUidTxBytes(int uid)：获取某个网络 UID 的发送字节数。

特殊方法及说明：

- 总接受流量：TrafficStats.getTotalRxBytes()；
- 总发送流量：TrafficStats.getTotalTxBytes()；
- 不包含 WIFI 的手机 GPRS 接收量：TrafficStats.getMobileRxBytes()；
- 不包含 Wifi 的手机 GPRS 发送量：TrafficStats.getMobileTxBytes()；
- 某一个进程的总接收量：TrafficStats.getUidRxBytes(Uid)；
- 某一个进程的总发送量：TrafficStats.getUidTxBytes(Uid)。

这些都是应用程序从第一次启动到最后一次启动的统计量。而不是"从本次开机到本次关机的统计量"。

下面，我们来寻找 TrafficStats 对象读取的是 Linux 下的哪些文件对象（真机测试，关闭模拟器）。

（1）在 Docs 窗口中执行 adb shell 命令，此时进入手机的 Linux 系统终端。

（2）执行 ls –l 命令，列出手机根目录中的所有文件的详细信息（如图 5-16 所示），这里重点关注 proc 这个文件。

（3）执行 cd proc，来到 proc 文件目录。

（4）执行 ls –l 命令，列出 proc 目录下的所有文件的详细信息（如图 5-17 所示）这里我们重点关注 uid_stat 这个文件（user id status 的缩写，应用程序的状态信息，每个应用程序都对应一个状态信息）。

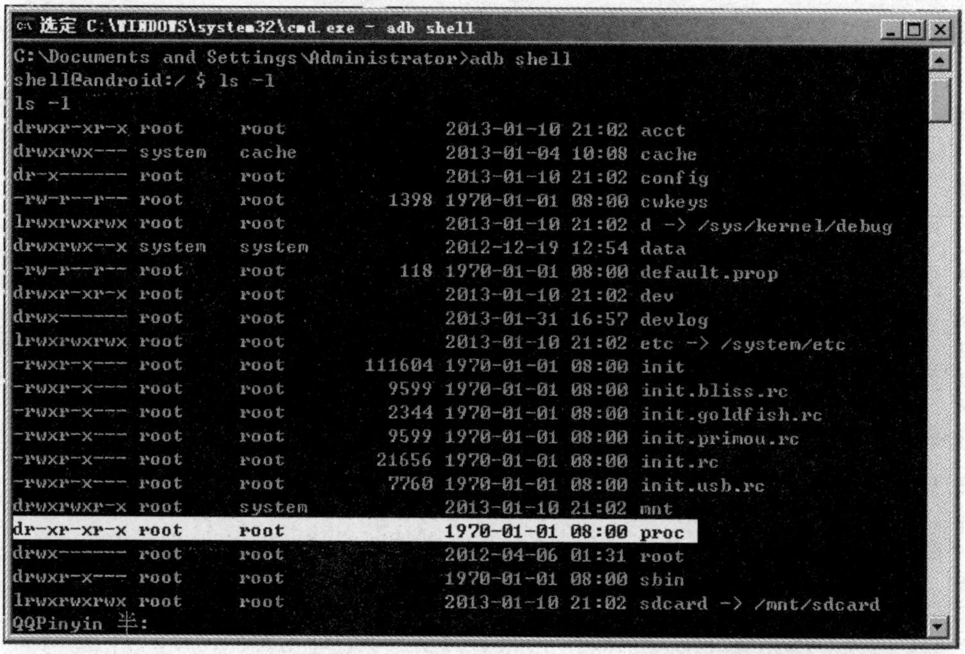

```
选定 C:\WINDOWS\system32\cmd.exe - adb shell
C:\Documents and Settings\Administrator>adb shell
shell@android:/ $ ls -l
ls -l
drwxr-xr-x root     root              2013-01-10 21:02 acct
drwxrwx--- system   cache             2013-01-04 10:08 cache
dr-x------ root     root              2013-01-10 21:02 config
-rw-r--r-- root     root         1398 1970-01-01 08:00 cwkeys
lrwxrwxrwx root     root              2013-01-10 21:02 d -> /sys/kernel/debug
drwxrwx--x system   system            2012-12-19 12:54 data
-rw-r--r-- root     root          118 1970-01-01 08:00 default.prop
drwxr-xr-x root     root              2013-01-10 21:02 dev
drwx------ root     root              2013-01-31 16:57 devlog
lrwxrwxrwx root     root              2013-01-10 21:02 etc -> /system/etc
-rwxr-x--- root     root       111604 1970-01-01 08:00 init
-rwxr-x--- root     root         9599 1970-01-01 08:00 init.bliss.rc
-rwxr-x--- root     root         2344 1970-01-01 08:00 init.goldfish.rc
-rwxr-x--- root     root         9599 1970-01-01 08:00 init.primou.rc
-rwxr-x--- root     root        21656 1970-01-01 08:00 init.rc
-rwxr-x--- root     root         7760 1970-01-01 08:00 init.usb.rc
drwxrwxr-x root     system            2013-01-10 21:02 mnt
dr-xr-xr-x root     root              1970-01-01 08:00 proc
drwx------ root     root              2012-04-06 01:31 root
drwxr-x--- root     root              1970-01-01 08:00 sbin
lrwxrwxrwx root     root              2013-01-10 21:02 sdcard -> /mnt/sdcard
QQPinyin 半:
```

图 5-16

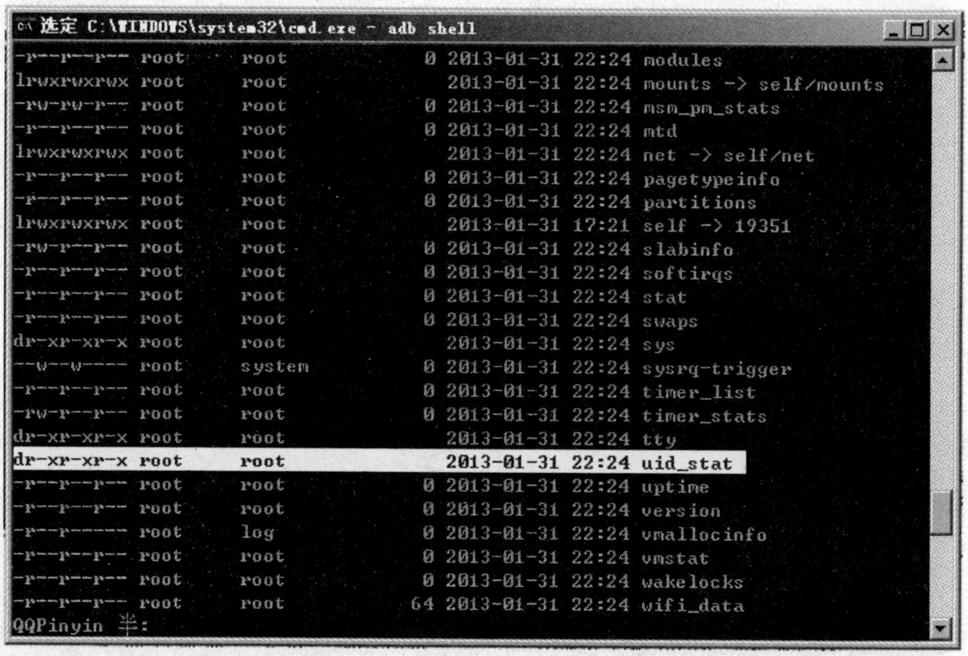

```
选定 C:\WINDOWS\system32\cmd.exe - adb shell
-r--r--r-- root     root            0 2013-01-31 22:24 modules
lrwxrwxrwx root     root              2013-01-31 22:24 mounts -> self/mounts
-rw-rw-r-- root     root            0 2013-01-31 22:24 msm_pm_stats
-r--r--r-- root     root            0 2013-01-31 22:24 mtd
lrwxrwxrwx root     root              2013-01-31 22:24 net -> self/net
-r--r--r-- root     root            0 2013-01-31 22:24 pagetypeinfo
-r--r--r-- root     root            0 2013-01-31 22:24 partitions
lrwxrwxrwx root     root              2013-01-31 17:21 self -> 19351
-rw-r--r-- root     root            0 2013-01-31 22:24 slabinfo
-r--r--r-- root     root            0 2013-01-31 22:24 softirqs
-r--r--r-- root     root            0 2013-01-31 22:24 stat
-r--r--r-- root     root            0 2013-01-31 22:24 swaps
dr-xr-xr-x root     root              2013-01-31 22:24 sys
--w--w---- root     system          0 2013-01-31 22:24 sysrq-trigger
-r--r--r-- root     root            0 2013-01-31 22:24 timer_list
-rw-r--r-- root     root            0 2013-01-31 22:24 timer_stats
dr-xr-xr-x root     root              2013-01-31 22:24 tty
dr-xr-xr-x root     root              2013-01-31 22:24 uid_stat
-r--r--r-- root     root            0 2013-01-31 22:24 uptime
-r--r--r-- root     root            0 2013-01-31 22:24 version
-r-------- root     log             0 2013-01-31 22:24 vmallocinfo
-r--r--r-- root     root            0 2013-01-31 22:24 vmstat
-r--r--r-- root     root            0 2013-01-31 22:24 wakelocks
-r--r--r-- root     root           64 2013-01-31 22:24 wifi_data
QQPinyin 半:
```

图 5-17

（5）执行 cd uid_stat 命令→回车后，执行 ls 命令（列出 uid_stat 目录下的所有文件）。

（6）执行 ls –l 命令来查看 uid_stat 目录下的文件详细信息（如图 5-18 所示），在该图中可以发现，每个条目的最左端都是以“d”开头的，说明该目录下的都是文件夹。每个条目最右端对应的数字其实都是文件夹对应的文件 id（例如，1000）。

（7）执行 cd 1000，此时进入了该文件中，回车后，执行 ls 命令会列出该目录下的所有文件信息，接着执行 ls –l 命令，此时显示出 1000 文件夹下所有文件的详细信息（如图 5-19 所示，其中，tcp_rcv 表示的是通过 TCP 协议接收到的数据，tcp_snd 表示的是通过 TCP 协议发送的数据）。

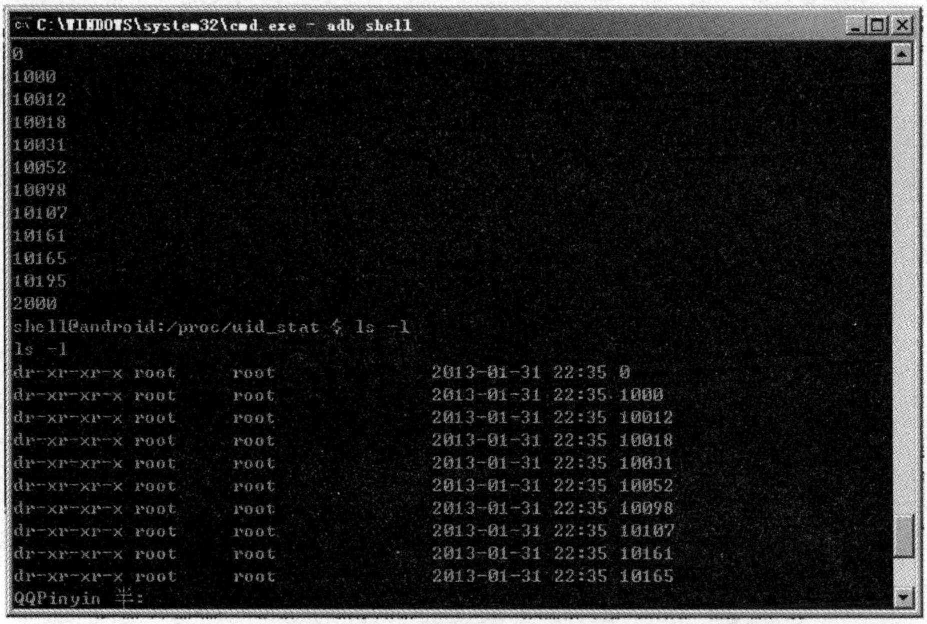

图 5-18

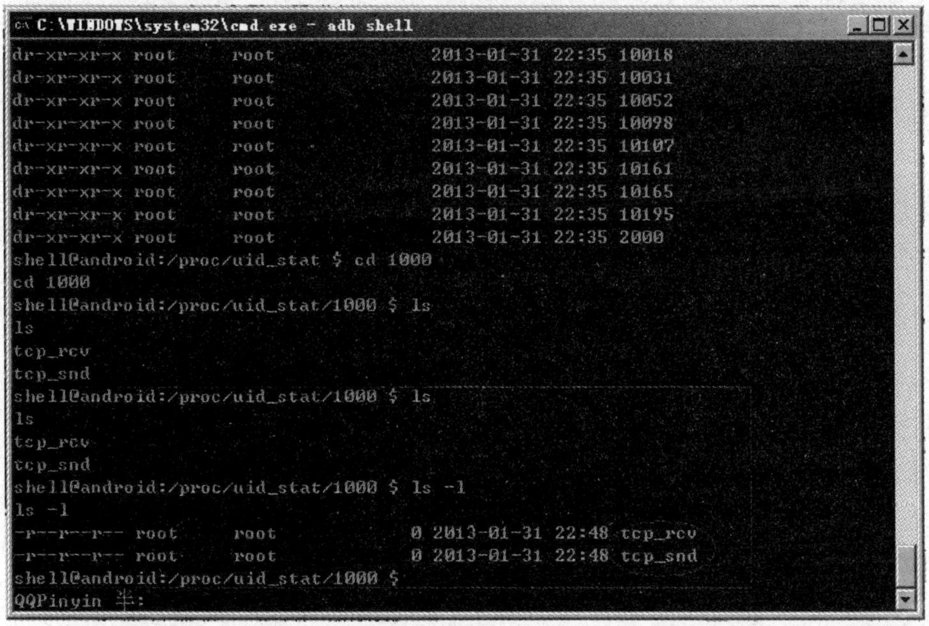

图 5-19

(8) 执行 cat tcp_rcv 查看接收到的字节数（对应的流量单位为 KB）、执行 cat tcp_snd 查看发送的字节数（对应的流量单位为 KB），执行结果如图 5-20 所示。

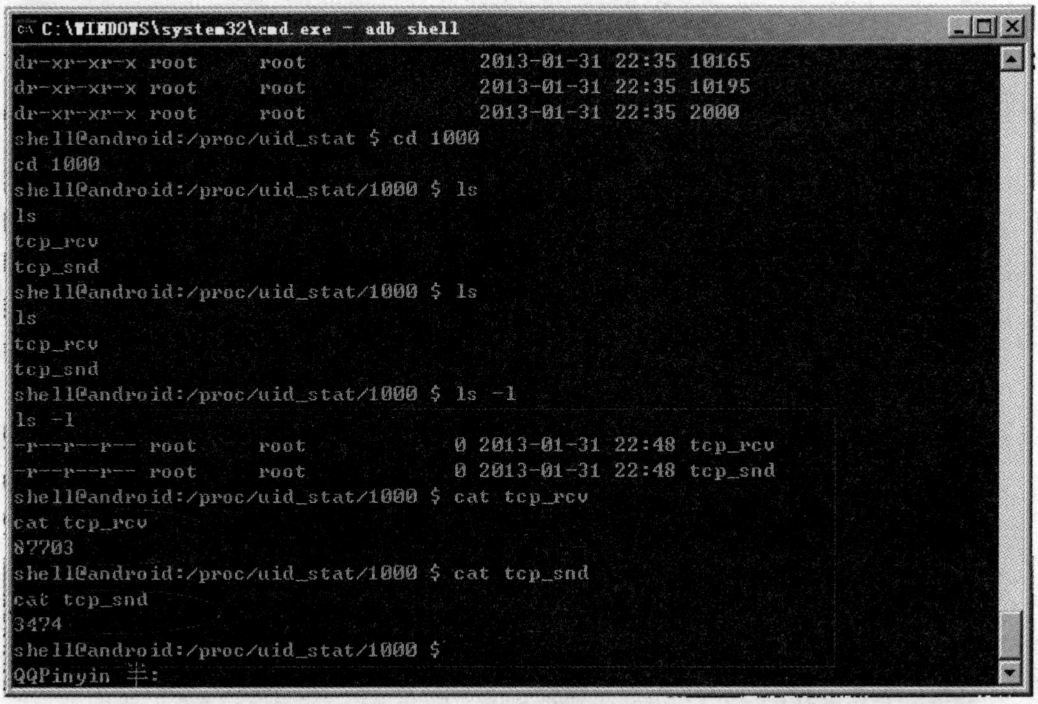

```
C:\WINDOWS\system32\cmd.exe - adb shell
dr-xr-xr-x root     root                2013-01-31 22:35 10165
dr-xr-xr-x root     root                2013-01-31 22:35 10195
dr-xr-xr-x root     root                2013-01-31 22:35 2000
shell@android:/proc/uid_stat $ cd 1000
cd 1000
shell@android:/proc/uid_stat/1000 $ ls
ls
tcp_rcv
tcp_snd
shell@android:/proc/uid_stat/1000 $ ls
ls
tcp_rcv
tcp_snd
shell@android:/proc/uid_stat/1000 $ ls -l
ls -l
-r--r--r-- root     root                0 2013-01-31 22:48 tcp_rcv
-r--r--r-- root     root                0 2013-01-31 22:48 tcp_snd
shell@android:/proc/uid_stat/1000 $ cat tcp_rcv
cat tcp_rcv
87703
shell@android:/proc/uid_stat/1000 $ cat tcp_snd
cat tcp_snd
3474
shell@android:/proc/uid_stat/1000 $
QQPinyin 半:
```

图 5-20

说明：在 Android 系统下，系统都会给每一个应用程序分配一个 uid(user id)和 gid (group id)。一般来说 uid 和 gid 是自动增长的，且两者都相同。每安装一个新的应用程序，系统都会分配 gid 和 uid 给新的应用程序；每卸载一个应用后，该应用的 uid 和 gid 会被系统回收，当再安装一个应用时，被卸载掉的应用所对应的 gid 和 uid 会分配给新安装的应用程序。我们进入到手机的 data\data 目录下查看该目录下的所有文件的详细信息（该目录下存放所有安装的应用程序，每安装一个应用程序，就会在该目录下生成一个与之对应的文件夹）。手机没有获取 Root 权限是不能够查看该目录下的文件（权限拒绝）的，此时我们通过模拟器来查看。执行 cd data/data 后回车进入到 data\data 对应的目录下，执行 ls –l 命令后如图 5-21 所示（该图所对应的第二列和第三列分别为 uid、gid，且两者相同。另外，每个应用程序也都会在 uid_stat 文件夹下生成一个与之对应的应用程序的 id 号）。通过观察可以得知 uid 和 uid_stat 文件夹下对应的应用程序的 id 号的关系是 u0_a26→10026，u0_a115→10115。

图 5-20 对应 uid_stat 下的 1000，而 1000 对应的应用应该是 u0_a1000，我们可以通过 u0_a1000 查看到该应用的名称，然后通过第三方的应用来检测其上传、下载的流量，最后与图 5-20 中的数字进行比较（注意单位的转换）。

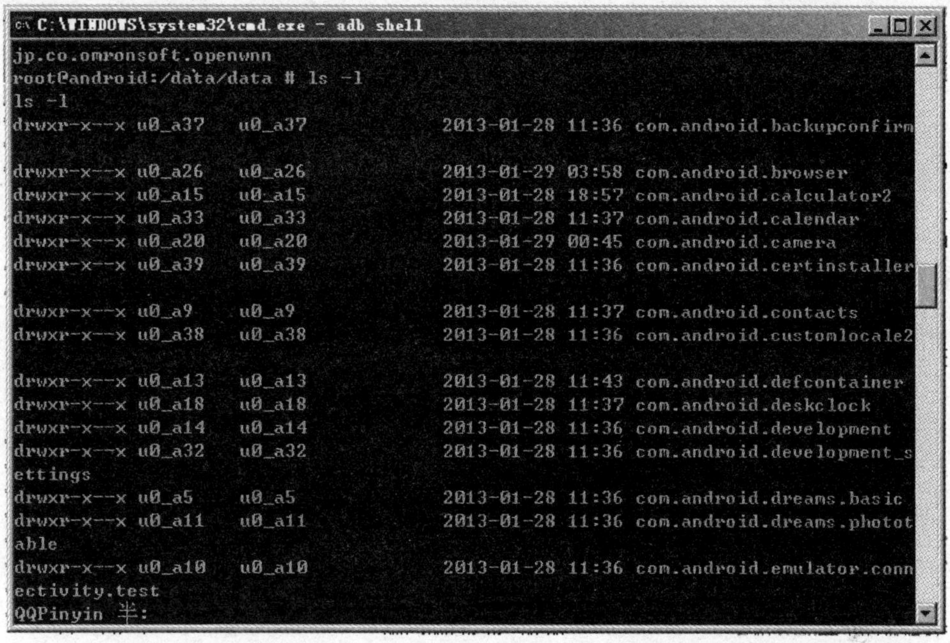

图 5-21

5.3.2 流量统计的实现

实现思路：通过 PackageManager 获取到所有具有 Intenet 权限信息的应用程序，通过应用的 uid 来分别获取到该应用程序的上传流量（TrafficStats.getUidTxBytes(uid)）、下载流量（TrafficStats.getUidRxBytes(uid)）。

流量统计对应的 Activity 为 TrafficInfoActivity，TrafficInfoActivity.java 的业务代码如下：

```
package com.guoshisp.mobilesafe;
import java.util.List;
import android.app.Activity;
import android.os.Bundle;
import android.os.Handler;
import android.text.format.Formatter;
import android.view.View;
import android.view.ViewGroup;
import android.widget.BaseAdapter;
import android.widget.ImageView;
import android.widget.LinearLayout;
import android.widget.ListView;
import android.widget.TextView;
import com.guoshisp.mobilesafe.domain.TrafficInfo;
import com.guoshisp.mobilesafe.engine.TrafficInfoProvider;
public class TrafficInfoActivity extends Activity {
    //展示数据列表
```

```java
private ListView lv;
//获取到所有具有Intenet权限的应用的流量信息
private TrafficInfoProvider provider;
//ProgressBar和TextView（正在加载...）的父控件，用于控制其显示
private LinearLayout ll_loading;
//封装单个具有Intenet权限的应用的流量信息
private List<TrafficInfo> trafficInfos;
//处理子线程发送过来的消息，更新UI
private Handler handler = new Handler(){
    public void handleMessage(android.os.Message msg) {
        ll_loading.setVisibility(View.INVISIBLE);
        lv.setAdapter(new TrafficAdapter());
    };
};
@Override
protected void onCreate(Bundle savedInstanceState) {
    setContentView(R.layout.traffic_info);
    super.onCreate(savedInstanceState);
    lv = (ListView) findViewById(R.id.lv_traffic_manager);
    provider = new TrafficInfoProvider(this);
    ll_loading = (LinearLayout) findViewById(R.id.ll_loading);
    ll_loading.setVisibility(View.VISIBLE);
    //获取到具有Internet权限的应用所产生的流量
    new Thread(){
        public void run() {
            trafficInfos = provider.getTrafficInfos();
            //向主线程中发送一个空消息，用于通知主线程更新数据
            handler.sendEmptyMessage(0);
        };
    }.start();
}
//数据适配器
private class TrafficAdapter extends BaseAdapter{
    public int getCount() {
        return trafficInfos.size();
    }
    public Object getItem(int position) {
        return trafficInfos.get(position);
    }
    public long getItemId(int position) {
        return position;
    }
    //ListView中显示多少个Item,该方法就被调用多少次
    public View getView(int position, View convertView, ViewGroup
                        parent) {
        View view;
```

```java
        ViewHolder holder = new ViewHolder();
        TrafficInfo info = trafficInfos.get(position);
        //复用缓存的View
        if(convertView==null){
            view = View.inflate(getApplicationContext(), R.layout.
                    traffic_item, null);
            holder.iv_icon = (ImageView) view.findViewById(R.id.iv_
                        traffic_icon);
            holder.tv_name = (TextView) view.findViewById(R.id.tv_
                        traffic_name);
            holder.tv_rx = (TextView) view.findViewById(R.id.tv_
                        traffic_rx);
            holder.tv_tx = (TextView) view.findViewById(R.id.tv_
                        traffic_tx);
            holder.tv_total = (TextView) view.findViewById(R.id.tv_
                        traffic_total);
            view.setTag(holder);
        }else{
            view = convertView;
            holder = (ViewHolder) view.getTag();
        }
        holder.iv_icon.setImageDrawable(info.getIcon());
        holder.tv_name.setText(info.getAppname());
        //下载所产生的流量
        long rx = info.getRx();
        //上传所产生的流量
        long tx = info.getTx();
        //增强程序的健壮性。因为在模拟器上运行时返回值为-1
        if(rx<0){
            rx = 0;
        }
        if(tx<0){
            tx = 0;
        }
        holder.tv_rx.setText(Formatter.formatFileSize
                        (getApplicationContext(), rx));
        holder.tv_tx.setText(Formatter.formatFileSize
                        (getApplicationContext(), tx));
        //总流量
        long total = rx + tx;
//通过Formatter将long类型的数据转换为MB或KB，当数字较小时，自动采用KB
        holder.tv_total.setText(Formatter.formatFileSize
                        (getApplicationContext(), total));
        return view;
    }
}
```

```java
//通过static的修饰，保证了栈内存中存在唯一一份字节码且被共用
static class ViewHolder{
    ImageView iv_icon;
    TextView tv_name;
    TextView tv_tx;
    TextView tv_rx;
    TextView tv_total;
}
}
```

该 Activity 对应的布局文件 traffic_info.xml 内容如下：

```xml
<?xml version="1.0" encoding="utf-8"?>
<LinearLayout xmlns:android="http://schemas.android.com/apk/res/android"
    android:layout_width="match_parent"
    android:layout_height="match_parent"
    android:background="@color/white"
    android:orientation="vertical" >
    <TextView
        style="@style/text_title_style"
        android:gravity="center"
        android:text="流量统计" />
    <SlidingDrawer
        android:layout_width="match_parent"
        android:layout_height="match_parent"
        android:content="@+id/content"
        android:handle="@+id/handle" >
        <ImageView
            android:id="@id/handle"
            android:layout_width="wrap_content"
            android:layout_height="wrap_content"
            android:src="@drawable/notification" />
        <FrameLayout
            android:id="@id/content"
            android:layout_width="match_parent"
            android:layout_height="match_parent" >
            <LinearLayout
                android:id="@+id/ll_loading"
                android:layout_width="match_parent"
                android:layout_height="match_parent"
                android:gravity="center"
                android:orientation="vertical"
                android:visibility="invisible" >
                <ProgressBar
                    style="@style/my_pb_style"
                    android:layout_width="wrap_content"
                    android:layout_height="wrap_content" />
```

```xml
            <TextView
                android:layout_width="wrap_content"
                android:layout_height="wrap_content"
                android:text="正在加载流量信息" />
        </LinearLayout>
        <ListView
            android:id="@+id/lv_traffic_manager"
            android:layout_width="match_parent"
            android:layout_height="match_parent" >
        </ListView>
    </FrameLayout>
</SlidingDrawer>
</LinearLayout>
```

文件解析：

（1）</SlidingDrawer>是一个类似于抽屉的控件，android:handle="@+id/handle"用于指定显示该控件的把手（这里显示一个小机器人图标），android:content="@+ id/content"用于指定显示该控件的内容（这里显示 ListView）。

（2）</ProgressBar>是自定义的 ProgressBar，style="@style/my_pb_style"用于指定自定义的样式（样式应该存放在 res/values/目录下），style.xml 文件最终内容如下：

```xml
<?xml version="1.0" encoding="utf-8"?>
<resources>
    <style name="text_title_style">
        <item name="android:layout_width">match_parent</item>
        <item name="android:layout_height">wrap_content</item>
        <item name="android:textColor">#66ff00</item>
        <item name="android:textSize">28sp</item>
    </style>
    <style name="image_divideline_style">
        <item name="android:layout_width">fill_parent</item>
        <item name="android:layout_height">1dip</item>
        <item name="android:layout_marginTop">5dip</item>
        <item name="android:background">@drawable/devide_line</item>
    </style>
    <style name="image_start_style">
        <item name="android:layout_width">wrap_content</item>
        <item name="android:layout_height">wrap_content</item>
        <item name="android:src">@android:drawable/star_big_on</item>
    </style>
    <style name="image_online_style">
        <item name="android:layout_width">wrap_content</item>
        <item name="android:layout_height">wrap_content</item>
        <item name="android:paddingLeft">3dip</item>
        <item name="android:src">@android:drawable/presence_online</item>
```

```xml
        </style>
        <style name="image_offline_style">
            <item name="android:paddingLeft">3dip</item>
            <item name="android:layout_width">wrap_content</item>
            <item name="android:layout_height">wrap_content</item>
            <item name="android:src">@android:drawable/presence_invisible</item>
        </style>
        <style name="image_logo_style">
            <item name="android:layout_width">fill_parent</item>
            <item name="android:layout_height">fill_parent</item>
            <item name="android:scaleType">center</item>
        </style>
        <style name="button_next_style">
            <item name="android:layout_width">wrap_content</item>
            <item name="android:layout_height">wrap_content</item>
            <item name="android:layout_alignParentBottom">true</item>
            <item name="android:layout_alignParentRight">true</item>
            <item name="android:text">下一步</item>
            <item name="android:drawableRight">@drawable/next</item>
            <item name="android:onClick">next</item>
        </style>
        <style name="button_pre_style">
            <item name="android:layout_width">wrap_content</item>
            <item name="android:layout_height">wrap_content</item>
            <item name="android:layout_alignParentBottom">true</item>
            <item name="android:layout_alignParentLeft">true</item>
            <item name="android:text">上一步</item>
            <item name="android:onClick">pre</item>
            <item name="android:drawableLeft">@drawable/previous</item>
        </style>
        <style name="text_content_style" parent="@style/text_title_style">
            <item name="android:textSize">20sp</item>
        </style>
        <style name="my_pb_style" parent="@android:style/Widget.ProgressBar">
            <item name="android:indeterminateDrawable">@drawable/my_pb_bg
                </item>
        </style>
</resources>
```

我们定义的 ProgressBar 的样式为：

```xml
<style name="my_pb_style" parent="@android:style/Widget.ProgressBar">
    <item name="android:indeterminateDrawable">@drawable/my_pb_bg</item>
</style>
```

可以看出，该 ProgressBar 继承了父样式（原始旋转的样式），我们只是覆盖了其中的一个属性：android:indeterminateDrawable，该属性用于显示 ProgressBar 的界面图形。my_pb_bg.xml

位于 drawable 目录下，该文件内容如下：

```xml
<?xml version="1.0" encoding="utf-8"?>
<animated-rotate xmlns:android="http://schemas.android.com/apk/res/android"
    android:drawable="@drawable/notification"
    android:pivotX="50%"
    android:pivotY="50%"
 />
```

其中的 android:drawable="@drawable/notification" 就是定义了 ProgressBar 在显示时的图形，android:pivotX="50%" 与 android:pivotY="50%" 决定了 ProgressBar 在旋转时的旋转重心（这里是以自身旋转）。

代码解析：

private TrafficInfoProvider provider 为获取到所有具有 Intenet 权限的应用的流量信息，该对象位于包 "com.guoshisp.mobilesafe.engine" 下，TrafficInfoProvider.java 中的代码如下：

```java
package com.guoshisp.mobilesafe.engine;
import java.util.ArrayList;
import java.util.List;
import android.content.Context;
import android.content.pm.PackageInfo;
import android.content.pm.PackageManager;
import android.net.TrafficStats;
import com.guoshisp.mobilesafe.domain.TrafficInfo;
public class TrafficInfoProvider {
    private PackageManager pm;
    private Context context;
    public TrafficInfoProvider(Context context) {
        this.context = context;
        pm = context.getPackageManager();
    }
    /**
     * 返回所有的有互联网访问权限的应用程序的流量信息
     * @return
     */
    public List<TrafficInfo> getTrafficInfos() {
        //获取到配置权限信息的应用程序
        List<PackageInfo> packinfos = pm
                .getInstalledPackages(PackageManager.GET_PERMISSIONS);
        //存放具有 Internet 权限信息的应用
        List<TrafficInfo> trafficInfos = new ArrayList<TrafficInfo>();
        for(PackageInfo packinfo : packinfos){
            //获取该应用的所有权限信息
            String[] permissions = packinfo.requestedPermissions;
            if(permissions!=null&&permissions.length>0){
                for(String permission : permissions){
```

```
                //筛选出具有Internet权限的应用程序
                if("android.permission.INTERNET".equals(permission)){
                    //用于封装具有Internet权限的应用程序信息
                    TrafficInfo trafficInfo = new TrafficInfo();
                    //封装应用信息
                    trafficInfo.setPackname(packinfo.packageName);
            trafficInfo.setIcon(packinfo.applicationInfo.loadIcon(pm));
    trafficInfo.setAppname(packinfo.applicationInfo.loadLabel(pm).
                    toString());
                    //获取到应用的uid(user id)
                    int uid = packinfo.applicationInfo.uid;
                    //TrafficStats对象通过应用的uid来获取应用的下载、上传流
                      量信息
    trafficInfo.setRx(TrafficStats.getUidRxBytes(uid));
    trafficInfo.setTx(TrafficStats.getUidTxBytes(uid));
                    trafficInfos.add(trafficInfo);
                    trafficInfo = null;
                    break;
                }
            }
        }
    }
    return trafficInfos;
    }
}
```

代码解析：

（1）在迭代应用权限，并筛选出具有 Internet 权限信息的应用时使用到了 break 语句。这样可以避免应用中存在两个或多个 Internet 权限出现的重复添加应用信息的情况发生。break 语句的作用是跳出当前循环体，不再执行当前循环中 break 下面的语句。Continue 是直接结束本次循环，进入下一次循环。

（2）TrafficInfo trafficInfo = new TrafficInfo()

用于封装具有 Internet 权限的应用程序信息，该对象位于包"com.guoshisp. mobilesafe. domain"下，TrafficInfo.java 代码如下：

```
package com.guoshisp.mobilesafe.domain;
import android.graphics.drawable.Drawable;
public class TrafficInfo {
    //应用的包名
    private String packname;
    //应用的名称
    private String appname;
    //上传的数据
    private long tx;
    //下载的数据
```

```java
    private long rx;
    //应用图标
    private Drawable icon;
    public String getPackname() {
        return packname;
    }
    public void setPackname(String packname) {
        this.packname = packname;
    }
    public String getAppname() {
        return appname;
    }
    public void setAppname(String appname) {
        this.appname = appname;
    }
    public long getTx() {
        return tx;
    }
    public void setTx(long tx) {
        this.tx = tx;
    }
    public long getRx() {
        return rx;
    }
    public void setRx(long rx) {
        this.rx = rx;
    }
    public Drawable getIcon() {
        return icon;
    }
    public void setIcon(Drawable icon) {
        this.icon = icon;
    }
}
```

在应用程序的主界面为"流量统计"添加单击事件，此时，MainActivity.java 的业务代码如下：

```java
package com.guoshisp.mobilesafe;
import android.app.Activity;
import android.content.Intent;
import android.os.Bundle;
import android.view.View;
import android.widget.AdapterView;
import android.widget.AdapterView.OnItemClickListener;
import android.widget.GridView;
import com.guoshisp.mobilesafe.adapter.MainAdapter;
```

```java
public class MainActivity extends Activity {
    //显示主界面中九大模块的GridView
    private GridView gv_main;
    @Override
    protected void onCreate(Bundle savedInstanceState) {
        super.onCreate(savedInstanceState);
        setContentView(R.layout.main);
        gv_main = (GridView) findViewById(R.id.gv_main);
        //为gv_main对象设置一个适配器，该适配器的作用是为每个item填充对应的数据
        gv_main.setAdapter(new MainAdapter(this));
        //为GridView对象中的item设置单击时的监听事件
        gv_main.setOnItemClickListener(new OnItemClickListener() {
            //参数一：item的父控件，也就是GridView；参数二：当前单击的item；参
            //  数三：当前单击的item在GridView中的位置
            //参数四：id的值为单击了GridView的哪一项对应的数值，单击GridView
            //  第9项，那id就等于8
            public void onItemClick(AdapterView<?> parent, View view,
                    int position, long id) {
                switch (position) {
                case 0: //手机防盗
                    //跳转到"手机防盗"对应的Activity界面
                    Intent lostprotectedIntent = new Intent(MainActivity.
                            this,LostProtectedActivity.class);
                    startActivity(lostprotectedIntent);
                    break;
                case 1: //通信卫士
                    Intent callSmsIntent = new Intent(MainActivity.this,
                            CallSmsSafeActivity.class);
                    startActivity(callSmsIntent);
                    break;
                case 2: //程序管理
                    Intent appManagerIntent = new Intent(MainActivity.
                            this,AppManagerActivity.class);
                    startActivity(appManagerIntent);
                case 3: //进程管理
                    Intent taskManagerIntent = new Intent(MainActivity.
                            this,TaskManagerActivity.class);
                    startActivity(taskManagerIntent);
                    break;
                case 4: //流量管理
                    Intent trafficInfoIntent = new Intent(MainActivity.
                            this,TrafficInfoActivity.class);
                    startActivity(trafficInfoIntent);
                    break;
                case 7://高级工具
```

```
                Intent atoolsIntent = new Intent(MainActivity.this,
                                    AtoolsActivity.class);
                startActivity(atoolsIntent);
                break;
            case 8://设置中心
                //跳转到"设置中心"对应的Activity界面
                Intent settingIntent = new Intent(MainActivity.this,
                                    SettingCenterActivity.class);
                startActivity(settingIntent);
                break;
            }
        }
    });
  }
}
```

测试运行：为组件配置好组件信息后运行。进入"流量管理"的主界面，如图 5-22 所示（屏幕底部有个把手），将把手拉上来后的界面如图 5-23 所示。

图 5-22   图 5-23

需要掌握的知识点小结

（1）使用 ActivityManager 获取具有权限信息的应用。

（2）ListView 的优化。

（3）自定义 ProgressBar。

（4）使用</SlidingDrawer>控件实现抽屉效果。

（5）使用 TrafficInfo 对象获取应用的上传、下载流量的信息。

（6）使用 Handler+Message 实现子线程与主线程的通信。

5.4 手机杀毒模块的设计

需求：精确扫描手机中已安装的应用程序是否含有病毒。

手机杀毒的 UI 如图 5-24 所示。

图 5-24

5.4.1 病毒查杀的原理

病毒的概念：二进制代码（一段恶意程序）。

下面介绍一般市场上的病毒查杀软件的执行原理：

（1）Windows 下的查杀原理（瑞星、金山、江民等杀毒软件）：最初查杀病毒都是查杀已知的病毒，提取出文件的特征码（例如，注册开机启动，往系统分区下释放文件）。之前的特征码都是存放在数据库中，将提取出来的特征码和数据库中的进行比对，如果存在就判断为病毒。这种传统的基于特征码的查杀方式到 2006 年出现了一个问题：随着病毒的越来越多，需要存入数据库中的特征码也越来越多，数据库体积倍增。当时的卡巴斯基或瑞星的特征码的体积已经有 70~80 MB 了，而下载这些软件时，需要将病毒库一同下载下来才可实现杀毒；卡巴斯基在 Windows 下已经存在 3000 万条特征码了，如果需要对计算机进行全盘扫描，需要扫描每个文件中的特征码，然后在这 3000 万条中比对一次，就算是 Oracle 也需要 7～8 秒，更何况是杀毒软件中内置的数据库，查询速度就更慢了，如果全盘扫描，一天也不够！

（2）升级的病毒查杀原理：杀毒引擎（优化的查询算法：不查询所有的文件，例如.txt 文件，会提取.exe 和.dll 文件的特征码，这样的文件存在病毒的可能性比较大，然后采用分批查询的方式查询数据库。所谓的多引擎就是多套的查询算法，将特征码快速定位到数据库中的数据）。

（3）目前的杀毒原理：云查杀。将常见的病毒特征码存到客户机上，不常见的保存在服务器上，当有一些可疑的文件出现时，它会将该文件的特征码提交到服务器上，由服务器来处理。而服务器的处理能力是比较强的，会将处理结果返回给客户端。

（4）互联网云安全计划：在安装杀毒软件时提示是否愿意加入，然后它自动扫描客户端，自动提交可疑文件，将扫描到的可疑文件直接移除，这是比较危险的（侵犯了用户的隐私）。

（5）主动防御（最新杀毒，监视系统关键的 API 是否改变，或者是注册表的操作）：360 经常提示我们 xx 程序正在加载，是否允许更改浏览器主页。

（6）人工智能：也是主动防御的一种。主要是用到了一套复杂的 if 语句。还用到了"模糊逻辑"：if...else 判断（传统的是 0 和 1 进行真假判断）。原理是不对数据进行真假的判断，而是进行一个范围内的取值。例如，Windows 下的病毒，在 run 节点添加启动项，在 C 盘的 system 下释放一些.exe 文件，或感染一些.exe 文件，如果它符合这其中的 2 条或 3 条，就会拦截它，在桌面的右下角给出提示（当前文件比较可疑，是否提交到服务器上）。这就是主动防御。

如何获取病毒的特征码：在电信运营商的主节点上部署服务器集群——蜜罐（它们没有打任何的补丁，也没有防火墙，主动让那些病毒来感染，然后提取这些病毒，通过静态或动态分析来提取它的特征码，最终将这些特征码存入数据库中）。采用的是主动捕获的方式获取的。

2012 年年底，Android 下的病毒已接近 26 000 款。所以，通过获取应用程序的特征码，并查询本地数据库中是否包含该特征码的做法是可行的。

5.4.2 手机杀毒的具体实现方法

我们先观察一下存储病毒特征码的病毒数据库 antivirus.db。使用 SQLite Expert 打开后，进入表 datable，如图 5-25 所示。

由图 5-25 可以发现，病毒的特征码是表中的 MD5 值，所以，只需要提取到应用应用程序的特征码，然后查询数据库，对比即可实现病毒的查询。杀毒的过程就是一个卸载应用程序的过程，需要将病毒存入缓存中，然后通过遍历卸载即可。

RecNo	_id	MD5	type	name	desc
		Click here to define a filter			
989	2044	644e30f28bd7715abb36 6d08f5921ebe	6	Android.Adware.AirAD.a	恶意后台扣费,病毒木马程序
990	2045	6eb5bb1d02dd9534f923 f457b7f0898e	6	Android.Troj.GacBlocker.b	恶意后台扣费,病毒木马程序
991	2046	898da8221558c4ddb38a 53111c41984f	6	Android.Troj.AirAD.a	恶意后台扣费,病毒木马程序
992	2047	1002f6c9087d1cecc983 311e2f27c7a0	6	Android.Hack.i22hkt.	恶意后台扣费,病毒木马程序
993	2048	1db87cca2666ae82f787 819fc68100ce	6	Android.Troj.AirAD.a	恶意后台扣费,病毒木马程序
994	2049	6f10b32b03fecb0089ca 3a780f31aa1a	6	Android.Adware.AirAD.a	恶意后台扣费,病毒木马程序
995	2050	d3eef588692616a325f3 f0a8c3cde5f9	6	Android.Hack.i22hkt.a	恶意后台扣费,病毒木马程序
996	2051	dfd3bd236e89b4cee455 1444930d730d	6	Android.Troj.DroidDream.b	恶意后台扣费,病毒木马程序
997	2052	3fac8c25fc65d8b4ae9c d72a49555294	6	Android.Adware.AirAD.a	恶意后台扣费,病毒木马程序
998	2053	bc0ad4560829dfbb364c 8d79a51f4eb8	6	Android.Troj.AdWooboo.a	恶意后台扣费,病毒木马程序
999	2054	cb53725327f0c591c117 ba494e8b004c	6	Android.Adware.AirAD.a	恶意后台扣费,病毒木马程序
1000	2055	feead50a562436d3e87b 15d59aa091e8	6	Android.Adware.AirAD.a	恶意后台扣费,病毒木马程序
1001	2056	8824c59dca432ed79ebc 7f6009d33a3d	6	Android.Hack.CarrierIQ.a	在未明示情况下窃取用户隐私
1002	2057	230fceebb6d9c801f3b2 aed58479a0a3	6	Android.Hack.CarrierIQ.a	在未明示情况下窃取用户隐私
1003	2058	527bd9d97219c0c9554b e6e7a062387a	6	Android.Hack.CarrierIQ.a	在未明示情况下窃取用户隐私
1004	2059	a49d672993ee261d94ec 1a074167d2fe	6	Android.Hack.CarrierIQ.a	在未明示情况下窃取用户隐私
1005	2060	838aad6362cd31a682d4 df22d4411c1b	6	Android.Hack.CarrierIQ.a	在未明示情况下窃取用户隐私
1006	2061	c5afa6d19590a0ba460d e9aceb718010	6	Android.Hack.CarrierIQ.a	在未明示情况下窃取用户隐私

图 5-25

首先,将病毒数据库文件 antivirus.db 复制到 assets 目录下,然后通过代码将其复制到手机系统中。这里在加载 Splash 界面时完成数据库的复制。SplashActivity.java 的业务代码见在线资源包中代码文本部分 5.4.2.doc。

代码解析:

```
new Thread() {
    public void run() {
        File file = new File(getFilesDir(), "antivirus.db");
        if (file.exists() && file.length() > 0) {//数据库文件已经复制成功
        } else {
            AssetCopyUtil.copy1(getApplicationContext(),
                "antivirus.db", file.getAbsolutePath(), null);
        }
    };
}.start();
```

将数据库复制到/data/data/com.guoshisp.mobilesafe/files 目录下。AssetCopyUtil.java 代码如下:

```
package com.guoshisp.mobilesafe.utils;
```

```java
import java.io.File;
import java.io.FileOutputStream;
import java.io.InputStream;
import java.io.OutputStream;
import android.app.ProgressDialog;
import android.content.Context;
import android.content.res.AssetManager;
/**
 * 资产文件复制的工具类
 *
 * @author
 *
 */
public class AssetCopyUtil {
    private Context context;
    public AssetCopyUtil(Context context) {
        this.context = context;
    }
    /**
     * 复制资产目录下的文件
     *
     * @param srcfilename
     * 源文件的名称
     * @param file
     * 目标文件的对象
     * @param pd
     * 进度条对话框
     * @return 是否复制成功
     */
    public boolean copyFile(String srcfilename, File file, ProgressDialog pd) {
        try {
            //获取到资产目录的管理器。因为数据库存放在该目录下
            AssetManager am = context.getAssets();
            //打开资产目录下的资源文件，获取一个输入流对象
            InputStream is = am.open(srcfilename);
            //获取到该文件的字节数
            int max = is.available();
            //设置进度条显示的最大进度
            pd.setMax(max);
            //创建一个输出流文件，用于接收输入流
            FileOutputStream fos = new FileOutputStream(file);
            //创建一个缓存区
            byte[] buffer = new byte[1024];
            int len = 0;
            //进度条的最开始的位置应该为0
            int process = 0;
```

```java
            while ((len = is.read(buffer)) != -1) {
                fos.write(buffer, 0, len);
                //让进度条不断地动态显示当前的复制进度
                process += len;
                pd.setProgress(process);
            }
            //刷新缓冲区，关流
            fos.flush();
            fos.close();
            return true;
        } catch (Exception e) {
            e.printStackTrace();
            return false;
        }
    }
    /**
     * 从资产目录复制文件
     *
     * @param context
     * @param filename
     *     资产目录的文件的名称
     * @param destfilename
     *     目标文件的路径
     * @return
     */
    public static File copy1(Context context, String filename,
            String destfilename, ProgressDialog pd) {
        try {
            InputStream in = context.getAssets().open(filename);
            int max = in.available();
            if (pd != null) {
                pd.setMax(max);
            }
            File file = new File(destfilename);
            OutputStream out = new FileOutputStream(file);
            byte[] byt = new byte[1024];
            int len = 0;
            int total = 0;
            while ((len = in.read(byt)) != -1) {
                out.write(byt, 0, len);
                total += len;
                if (pd != null) {
                    pd.setProgress(total);
                }
            }
```

```java
            out.flush();
            out.close();
            in.close();
            return file;
        } catch (Exception e) {
            e.printStackTrace();
            return null;
        }
    }
}
```

病毒查杀对应的 Activity 为 AntiVirusActivity，AntiVirusActivity.java 业务代码如下：

```java
package com.guoshisp.mobilesafe;
import java.util.ArrayList;
import java.util.List;
import java.util.Map;
import android.app.Activity;
import android.content.Intent;
import android.content.pm.PackageInfo;
import android.content.pm.PackageManager;
import android.net.Uri;
import android.os.Bundle;
import android.os.Handler;
import android.os.Message;
import android.view.View;
import android.view.animation.Animation;
import android.view.animation.RotateAnimation;
import android.widget.ImageView;
import android.widget.LinearLayout;
import android.widget.ProgressBar;
import android.widget.TextView;
import android.widget.Toast;
import com.guoshisp.mobilesafe.db.dao.AntiVirusDao;
import com.guoshisp.mobilesafe.utils.Md5Encoder;
public class AntiVirusActivity extends Activity {
    protected static final int SCAN_NOT_VIRUS = 90;
    protected static final int FIND_VIRUS = 91;
    protected static final int SCAN_FINISH = 92;
    //查杀病毒时，雷达上的扫描指针
    private ImageView iv_scan;
    //应用程序包管理器
    private PackageManager pm;
    //操作数据库的对象
    private AntiVirusDao dao;
    //扫描进度条
```

```java
        private ProgressBar progressBar1;
        //显示发现的病毒数目
        private TextView tv_scan_status;
        //显示扫描的程序信息
        private LinearLayout ll_scan_status;
        //用于添加扫描到的病毒信息
        private List<PackageInfo> virusPackInfos;
        //旋转动画
        RotateAnimation ra;
        //存放病毒的集合
        private Map<String, String> virusMap;
        //用于与子线程通信，更新主线程（UI 线程）
        private Handler handler = new Handler() {
            public void handleMessage(android.os.Message msg) {
                PackageInfo info = (PackageInfo) msg.obj;
                switch (msg.what) {
                case SCAN_NOT_VIRUS://未发现病毒
                    TextView tv = new TextView(getApplicationContext());
                    tv.setText("扫描" + info.applicationInfo.loadLabel(pm) + "
                            安全");
                    ll_scan_status.addView(tv, 0);//添加到ll_scan_info控件最上面
                    break;
                case FIND_VIRUS://发现病毒
                    //将病毒添加到集合中
                    virusPackInfos.add(info);
                    break;
                case SCAN_FINISH://扫描完成
                    //停止动画的播放
                    iv_scan.clearAnimation();
                    //判断病毒集合的大小
                    if (virusPackInfos.size() == 0) {
                        Toast.makeText(getApplicationContext(), "扫描完毕,你的
                                手机很安全", 0)
                                .show();
                    }
                    break;
                }
            };
        };
        @Override
        protected void onCreate(Bundle savedInstanceState) {
            setContentView(R.layout.anti_virus);
            pm = getPackageManager();
            dao = new AntiVirusDao(this);
```

```java
        virusPackInfos = new ArrayList<PackageInfo>();
        super.onCreate(savedInstanceState);
        tv_scan_status = (TextView) findViewById(R.id.tv_scan_status2);
        iv_scan = (ImageView) findViewById(R.id.iv_scan);
        //设置一个旋转的动画
        ra = new RotateAnimation(0, 360, Animation.RELATIVE_TO_SELF, 1.0f,
                Animation.RELATIVE_TO_SELF, 1.0f);
        ra.setDuration(1000);
        //设置旋转的重复次数（一直旋转）
        ra.setRepeatCount(Animation.INFINITE);
        //设置旋转的模式（旋转一个回合后，重新旋转）
        ra.setRepeatMode(Animation.RESTART);
        ll_scan_status = (LinearLayout) findViewById(R.id.ll_scan_status);
        progressBar1 = (ProgressBar) findViewById(R.id.progressBar1);
}
public void kill(View v) {
        //重置动画
        ra.reset();
        //启动动画
        iv_scan.startAnimation(ra);
        //开启一条子线程，遍历手机中各个应用的签名信息
        new Thread() {
            public void run() {
                //PackageManager.GET_SIGNATURES 应用程序的签名信息
                List<PackageInfo> packinfos = pm
                    .getInstalledPackages(PackageManager.GET_SIGNATURES);
                progressBar1.setMax(packinfos.size());
                //计数当前已经遍历了多少条应用程序，以显示查杀的进度
                int count = 0;
                //遍历出各个应用程序对应的签名信息
                for (PackageInfo info : packinfos) {
                    //将应用程序的签名信息转成 MD5 值，用于与病毒数据库比对
                    String md5 = Md5Encoder.encode(info.signatures[0]
                        .toCharsString());
                    //在病毒数据库中查找该 MD5 值，来判断该应用程序是否属于病毒
                    String result = dao.getVirusInfo(md5);
                    //如果查找的结果为 null，则表示当前遍历的应用不是病毒
                    if (result == null) {
                        Message msg = Message.obtain();
                        msg.what = SCAN_NOT_VIRUS;
                        msg.obj = info;
                        handler.sendMessage(msg);
                    } else {//当前遍历到的应用属于病毒
                        Message msg = Message.obtain();
```

```java
                msg.what = FIND_VIRUS;
                msg.obj = info;
                handler.sendMessage(msg);
            }
            count++;
            try {
                Thread.sleep(20);
            } catch (InterruptedException e) {
                e.printStackTrace();
            }
            progressBar1.setProgress(count);
        }
        //遍历结束
        Message msg = Message.obtain();
        msg.what = SCAN_FINISH;
        handler.sendMessage(msg);
    };
}.start();
}
// "一键清理"按钮
public void clean(View v) {
    //判断病毒集合的大小
    if (virusPackInfos.size() > 0) {
        for (PackageInfo pinfo : virusPackInfos) {
            //卸载应用程序
            String packname = pinfo.packageName;
            Intent intent = new Intent();
            intent.setAction(Intent.ACTION_DEFAULT);
            intent.setData(Uri.parse("package:" + packname));
            startActivity(intent);
        }
    }else{
        return;
    }
}
}
```

该界面对应的布局文件 anti_virus.xml 内容如下:

```xml
<?xml version="1.0" encoding="utf-8"?>
<LinearLayout xmlns:android="http://schemas.android.com/apk/res/android"
    android:layout_width="match_parent"
    android:layout_height="match_parent"
    android:orientation="vertical" >
    <LinearLayout
        android:layout_width="fill_parent"
```

```xml
        android:layout_height="40dp"
        android:background="#ff6cbd45" >
        <TextView
            android:id="@+id/iv_main_title"
            android:layout_width="fill_parent"
            android:layout_height="fill_parent"
            android:gravity="center"
            android:text="手 机 杀 毒"
            android:textColor="#ffffff"
            android:textSize="18sp" />
    </LinearLayout>
    <LinearLayout
        android:layout_width="match_parent"
        android:layout_height="wrap_content"
        android:orientation="horizontal" >
        <FrameLayout
            android:layout_width="100dip"
            android:layout_height="100dip" >
            <ImageView
                android:layout_width="100dip"
                android:layout_height="100dip"
                android:src="@drawable/ic_scanner_malware" />
            <ImageView
                android:id="@+id/iv_scan"
                android:layout_width="50dip"
                android:layout_height="50dip"
                android:src="@drawable/scan" />
        </FrameLayout>
        <LinearLayout
            android:layout_width="match_parent"
            android:layout_height="fill_parent"
            android:gravity="center_vertical"
            android:orientation="vertical" >
            <TextView
                style="@style/text_content_style"
                android:text="正在查杀" />
            <ProgressBar
                android:id="@+id/progressBar1"
                style="?android:attr/progressBarStyleHorizontal"
                android:layout_width="fill_parent"
                android:layout_height="wrap_content" />
        </LinearLayout>
    </LinearLayout>
    <RelativeLayout
        android:layout_width="fill_parent"
```

```xml
        android:layout_height="wrap_content" >
        <TextView
            android:id="@+id/tv_scan_status2"
            android:layout_width="wrap_content"
            android:layout_height="wrap_content"
            android:text="扫描状态"
            android:textSize="22sp" />
        <Button
            android:layout_width="wrap_content"
            android:layout_height="wrap_content"
            android:layout_alignParentRight="true"
            android:onClick="clean"
            android:paddingRight="5dip"
            android:text="一键清理" />
    </RelativeLayout>
    <ScrollView
        android:layout_width="fill_parent"
        android:layout_height="fill_parent"
        android:layout_weight="100"
        android:scrollbars="none" >
        <LinearLayout
            android:id="@+id/ll_scan_status"
            android:layout_width="fill_parent"
            android:layout_height="fill_parent"
            android:orientation="vertical" >
        </LinearLayout>
    </ScrollView>
    <LinearLayout
        android:layout_width="fill_parent"
        android:layout_height="wrap_content"
        android:orientation="horizontal" >
        <Button
            android:layout_width="fill_parent"
            android:layout_height="wrap_content"
            android:layout_marginLeft="3dp"
            android:layout_weight="1"
            android:background="@drawable/button_bg_selector"
            android:onClick="kill"
            android:text="查杀" />
    </LinearLayout>
</LinearLayout>
```

文件解析：android:src="@drawable/scan"中的 scan.xml 是定义的一个图形文件（雷达指针）。该文件位于 drawable 目录下，scan.xml 文件如下：

```xml
<?xml version="1.0" encoding="utf-8"?>
```

```xml
<shape xmlns:android="http://schemas.android.com/apk/res/android"
    android:shape="rectangle">
    <solid android:color="#66000000"/>
</shape>
```

代码解析：

（1）ra.setRepeatCount(Animation.INFINITE)与 ra.setRepeatMode(Animation. RESTART)可以保证旋转动画一直处于循环旋转状态，如果不给予设置，动画旋转一个回合后便停止播放。

（2）ra.reset()重置动画——让动画重新开始播放。

（3）private AntiVirusDao dao 用于操作病毒数据库的 AntiVirusDao 位于包 "com. guoshisp.mobilesafe.db.dao" 下，AntiVirusDao.java 代码如下：

```java
package com.guoshisp.mobilesafe.db.dao;
import android.content.Context;
import android.database.Cursor;
import android.database.sqlite.SQLiteDatabase;
public class AntiVirusDao {
    private Context  context;
    public AntiVirusDao(Context context) {
        this.context = context;
    }
    /**
     * 获取病毒的信息，如果没有获取到则返回空
     * @param md5
     * @return
     */
    public String getVirusInfo(String md5){
        //默认情况下没有获取到病毒信息
        String result = null;
        //病毒数据库的路径
        String path = "/data/data/com.guoshisp.mobilesafe/files/antivirus.db";
        //打开数据库
        SQLiteDatabase db = SQLiteDatabase.openDatabase(path, null,
                SQLiteDatabase.OPEN_READONLY);
        if(db.isOpen()){
            //执行查询操作，返回一个结果集
            Cursor cursor = db.rawQuery("select desc from datable where md5=?", new String[]{md5});
            if(cursor.moveToFirst()){
                result = cursor.getString(0);
            }
            //必须关闭系统的游标，如果没有关闭，即使关闭了数据库，也容易报出内存泄漏的异常信息
            cursor.close();
            db.close();
```

```
        }
        return result;
    }
}
```

另外,在进行病毒查杀的过程中,屏幕容易进行横、竖屏切换。默认情况下,横、竖屏切换是会重新调用生命周期方法的。为了避免此情况发生,需要在为 AntiVirusActivity 配置组件信息时,添加一些属性信息:

```
<activity android:name=".AntiVirusActivity"
          android:screenOrientation="portrait"
          android:configChanges="orientation|keyboardHidden">
</activity>
```

其中,android:screenOrientation="portrait"是设置该应用程序只能够竖屏显示;android:configChanges="orientation|keyboardHidde,当横、竖屏切换时会直接调用 onCreate 方法中的 onConfigurationChanged 方法,而不会重新执行 onCreate 方法。一般情况下,两者可以混合使用。

最后,需要在主界面中为"手机杀毒"设置一下单击事件,MainActivity.java 的业务代码如下:

```
package com.guoshisp.mobilesafe;
import android.app.Activity;
import android.content.Intent;
import android.os.Bundle;
import android.view.View;
import android.widget.AdapterView;
import android.widget.AdapterView.OnItemClickListener;
import android.widget.GridView;
import com.guoshisp.mobilesafe.adapter.MainAdapter;
public class MainActivity extends Activity {
    //显示主界面中九大模块的 GridView
    private GridView gv_main;
    @Override
    protected void onCreate(Bundle savedInstanceState) {
        super.onCreate(savedInstanceState);
        setContentView(R.layout.main);
        gv_main = (GridView) findViewById(R.id.gv_main);
//为 gv_main 对象设置一个适配器,该适配器的作用是用于为每个 item 填充对应的数据
        gv_main.setAdapter(new MainAdapter(this));
        //为 GridView 对象中的 item 设置单击时的监听事件
        gv_main.setOnItemClickListener(new OnItemClickListener() {
        //参数一:item 的父控件,也就是 GridView;参数二:当前单击的 item;参数三:
            当前单击的 item 在 GridView 中的位置
        //参数四:id 的值为单击 GridView 的哪一项对应的数值,单击 GridView 第 9
```

项，那id就等于8

```java
public void onItemClick(AdapterView<?> parent, View view,
        int position, long id) {
    switch (position) {
    case 0: //手机防盗
        //跳转到"手机防盗"对应的Activity界面
        Intent lostprotectedIntent = new Intent(MainActivity.
                            this, LostProtectedActivity.class);
        startActivity(lostprotectedIntent);
        break;
    case 1: //通信卫士
        Intent callSmsIntent = new Intent(MainActivity.this,
                            CallSmsSafeActivity.class);
        startActivity(callSmsIntent);
        break;
    case 2: //程序管理
        Intent appManagerIntent = new Intent(MainActivity.
                            this, AppManagerActivity.class);
        startActivity(appManagerIntent);
    case 3: //进程管理
        Intent taskManagerIntent = new Intent(MainActivity.
                            this, TaskManagerActivity.class);
        startActivity(taskManagerIntent);
        break;
    case 4: //流量管理
        Intent trafficInfoIntent = new Intent(MainActivity.
                            this, TrafficInfoActivity.class);
        startActivity(trafficInfoIntent);
        break;
    case 5: //手机杀毒
        Intent antiVirusIntent = new Intent(MainActivity.this,
                            AntiVirusActivity.class);
        startActivity(antiVirusIntent);
        break;
    case 7://高级工具
        Intent atoolsIntent = new Intent(MainActivity.this,
                            AtoolsActivity.class);
        startActivity(atoolsIntent);
        break;
    case 8://设置中心
        //跳转到"设置中心"对应的Activity界面
        Intent settingIntent = new Intent(MainActivity.this,
                            SettingCenterActivity.class);
        startActivity(settingIntent);
```

```
                    break;
                }
            }
        });
    }
}
```

测试运行：单击"手机杀毒"后，进入的界面如图 5-26 所示。单击"查杀"按钮后，界面如图 5-27 所示。查杀结束后的界面如图 5-28 所示。

图 5-26

图 5-27

图 5-28

 需要掌握的知识点小结

（1）在 Splash 界面中复制数据库。

（2）获取应用程序的特征码。

（3）控制旋转动画的循环播放。

（4）如何打开一个已经存在的数据库。

（5）使用 Handler+Message 实现主线程与子线程的通信。

（6）横、竖屏切换。

5.5 系统优化的功能介绍与 UI 设计

需求：清理手机中所有应用产生的缓存信息。

UI 视图如图 5-29 所示。

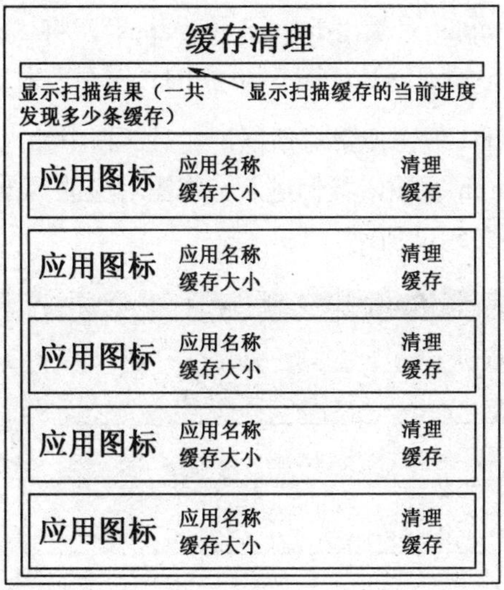

图 5-29

5.5.1 采用反射技术来调用系统隐藏的 API

在实现一些功能时,很多功能的实现过程不用我们自己重新去写,只需要查看一下系统的源代码就可以轻松地实现。在此,向读者介绍如何通过源码来实现缓存清理的需求。

进入模拟器中的设置→应用→手机安全卫士,此时可以看到该应用的缓存信息,如图 5-30 所示。

图 5-30

打开系统源代码，来到 apps 目录"dir\packages\apps"，将 Settings 文件夹复制到 Eclipse 中做为一个新的工程。我们想知道的是：图 5-30 中的"缓存"对应的数值是如何获取出来的。

在 Eclipse 中，使用 Ctrl+H 调出搜索对话框，在对话框中输入"缓存"（如图 5-31 所示），单击"search"后，弹出 Search 视图，我们进入该视图对应的 Settings 工程目录中，最终会进入到 strings.xml 文件中，如图 5-32 所示。

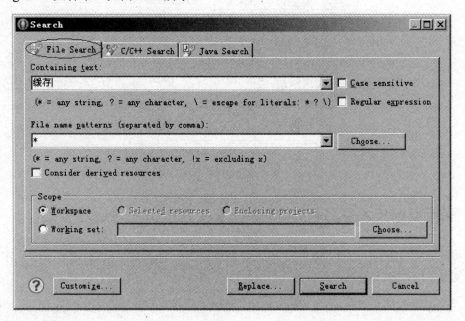

图 5-31

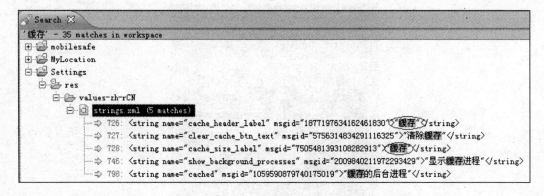

图 5-32

在该视图中，我们可以发现有两个条目与搜索吻合：<string name="cache_header_label"msgid="1877197634162461830">"缓存"</string>与<string name="cache_size_label" msgid="7505481393108282913">"缓存"</string>。由 name 属性可以知道：cache_header_label 表示的是图 5-30 中的标题（图 5-30 中有两个"缓存"字样，表示的是第一个），cache_size_label" msgid 表示的是图 5-30 中圈定的缓存。所以应该继续搜索"cache_size_label" msgid"在何处被应用到（双击该条目可以进入该条目对应的文件中）。搜索后，搜索结果直接定位到了"197:

android:text="@string/cache_size_label""条目,双击该条目后,进入到了该条目所对应的布局文件的预览图界面,在该预览图界面中,可以看到"Cache……"字样,在该字样的最右端有一个 TextView 控件(如图 5-33 所示)。

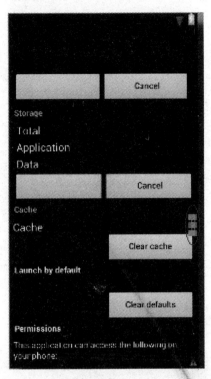

图 5-33

该控件用于显示缓存数据的大小,双击该控件即可进入该控件对应的 XML 文件(如图 5-34 所示),然后继续搜索该控件在什么时候被使用过(在搜索界面中搜索"cache_size_text"),此时可以定位到引用该 id 的源码的位置(如图 5-35 所示),双击"InstalledAppDetails.java"即可打开该源码,在源码中可以通过 Ctrl+F 搜索"cache_size_text"即可搜索到该 id 对应的变量名"mCacheSize",该对象是一个 TextView,我们下一步需要做的就是搜索到该变量在什么时候被赋值。继续通过 Ctrl+F 搜索"mCacheSize"即可定位到"mCacheSize.setText(getSizeStr(mAppEntry.cacheSize))",按住 Ctrl 键单击 getSizeStr 方法即可查看该方法,该方法对应的源码如图 5-36 所示。

```
215    <TextView
216        android:id="@+id/cache_size_text"
217        android:textAppearance="?android:attr/textAppearanceMedium"
218        android:paddingTop="6dip"
219        android:paddingRight="6dip"
220        android:layout_height="wrap_content"
221        android:layout_width="wrap_content"
222        android:maxLines="1" />
```

图 5-34

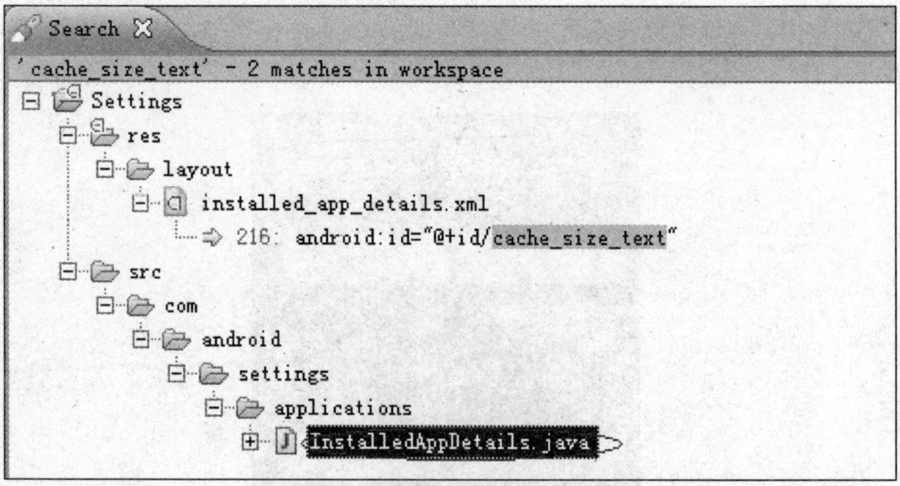

图 5-35

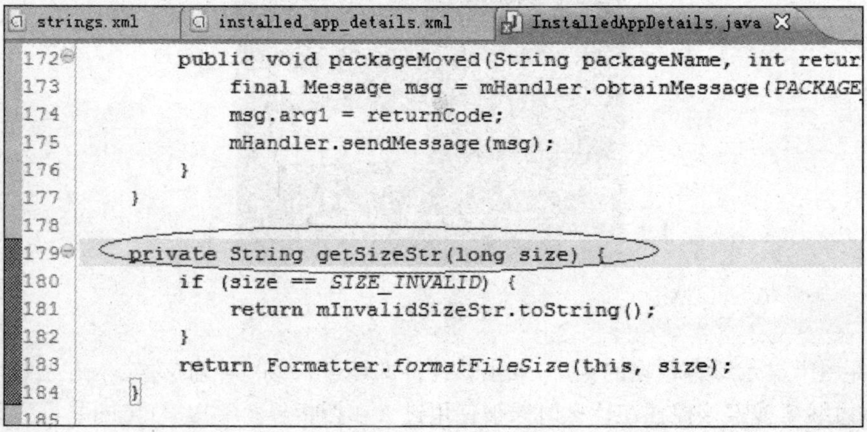

图 5-36

通过图 5-36 我们可以得知：该方法中传入的数值就是缓存的大小。继续通过 Ctrl+F 来搜索 mAppEntry 是何时被定义出来的，通过搜索，定位到"private ApplicationsState. AppEntry mAppEntry"上，但是 ApplicationsState 对象的全类名为"com.android.settings.applications. ApplicationsState"，并不是由系统提供的，所以需要跟踪一下 AppEntry（按住 Ctrl 键，鼠标左键单击 AppEntry）对象，跟踪结果如图 5-37 所示。由图 5-37 可知：AppEntry 继承了 SizeInfo，而在 SizeInfo 中定义有"long cacheSize"，所以只需要查看一下 cacheSize 在何时被赋值即可。继续通过通过 Ctrl+F 搜索 cacheSize，最终搜索到"entry.cacheSize = stats.cacheSize"，所以，需要跟踪 stats 对象（该对象的全类名为"android.content.pm. PackageStats"，这是系统提供的 jar 包，可以直接使用），查看该对象什么时候被赋值，跟踪结果如图 5-38 所示。

第5章 其他模块的设计

```
InstalledAppDetails.java    ApplicationsState.java
69
70    public static class SizeInfo {
71        long cacheSize;
72        long codeSize;
73        long dataSize;
74    }
75
76    public static class AppEntry extends SizeInfo {
77        final File apkFile;
78        final long id;
79        String label;
80        long size;
81
82        boolean mounted;
83
84        String getNormalizedLabel() {
85            if (normalizedLabel != null) {
86                return normalizedLabel;
87            }
88            normalizedLabel = normalize(label);
89            return normalizedLabel;
90        }
```

图 5-37

```
InstalledAppDetails.java    ApplicationsState.java
645    class BackgroundHandler extends Handler {
646        static final int MSG_REBUILD_LIST = 1;
647        static final int MSG_LOAD_ENTRIES = 2;
648        static final int MSG_LOAD_ICONS = 3;
649        static final int MSG_LOAD_SIZES = 4;
650
651        boolean mRunning;
652
653        final IPackageStatsObserver.Stub mStatsObserver = new IPackageStatsObserver.Stub() {
654            public void onGetStatsCompleted(PackageStats stats, boolean succeeded) {
655                boolean sizeChanged = false;
```

图 5-38

通过图 5-38 我们可以发现，该对象是一个接口中未实现的方法中的一个参数（final IPackageStatsObserver.Stub mStatsObserver = new IPackageStatsObserver.Stub()中的 Stub()就是未实现的接口），我们继续跟踪 mStatsObserver 接口被谁使用，最终，我们可以定位到 "mPm.getPackageSizeInfo(mCurComputingSizePkg, mStatsObserver)"，其中，mPm 就是系统的包管理者——PackageManager，getPackageSizeInfo 方法中的第一个参数表示的是包名，第二个参数就是跟踪的接口。

跟踪到这里，我们已经明白获取缓存的逻辑代码：通过 PackageManager 对象（系统中的 jar 包）来调用 getPackageSizeInfo(mCurComputingSizePkg, mStatsObserver)方法即可实现。

测试：通过 PackageManager 对象调用 getPackageSizeInfo(mCurComputingSizePkg, mStatsObserver)方法得到对应包名的缓存大小。

在测试的过程中，虽然我们得到了 PackageManager 对象，但是却不能够直接调用到 getPackageSizeInfo(mCurComputingSizePkg, mStatsObserver)方法。此时，查看 PackageManager

对象的源码，在源码中搜索该方法后，查看该方法说明得知：该方法被隐藏，如图5-39所示。

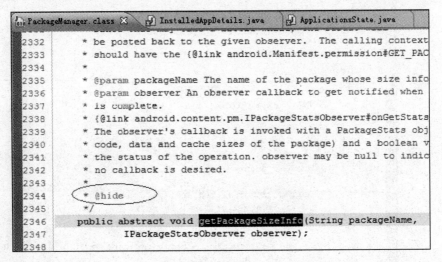

图 5-39

解决方案：该方法虽然被隐藏掉，但是我们可以通过反射技术来获取该方法。

5.5.2 系统优化的具体实现

实现思路：首先获取到手机中已安装的应用程序，然后遍历应用程序，接着获取到存在缓存的应用程序，最后为每个存在缓存的应用程序设置单击事件，该单击事件被触发后会跳转至清理缓存的界面。

系统优化对应的 Activity 为 CleanCacheActivity，CleanCacheActivity.java 中对应的业务代码如下：

```java
package com.guoshisp.mobilesafe;
import java.lang.reflect.Method;
import java.util.ArrayList;
import java.util.HashMap;
import java.util.List;
import java.util.Map;
import android.app.Activity;
import android.content.Intent;
import android.content.pm.IPackageStatsObserver;
import android.content.pm.PackageInfo;
import android.content.pm.PackageManager;
import android.content.pm.PackageStats;
import android.graphics.drawable.Drawable;
import android.net.Uri;
import android.os.AsyncTask;
import android.os.Build;
```

```java
import android.os.Bundle;
import android.os.RemoteException;
import android.text.format.Formatter;
import android.view.View;
import android.view.View.OnClickListener;
import android.widget.ImageView;
import android.widget.LinearLayout;
import android.widget.ProgressBar;
import android.widget.TextView;
public class CleanCacheActivity extends Activity {
    //显示扫描的进度
    private ProgressBar pd;
    //提示扫描的状态
    private TextView tv_clean_cache_status;
    //系统的包管理器
    private PackageManager pm;
    //存储带有缓存的应用的名称
    private List<String> cachePagenames;
    //显示所有带有缓存的应用程序信息
    private LinearLayout ll_clean;
    //存放缓存信息
    private Map<String, Long> cacheinfo;
    @Override
    protected void onCreate(Bundle savedInstanceState) {
        super.onCreate(savedInstanceState);
        setContentView(R.layout.ceanl_ce);
        pd = (ProgressBar) findViewById(R.id.progressBar1);
        ll_clean = (LinearLayout) findViewById(R.id.ll_clean_cache_cont);
        tv_clean_cache_status = (TextView) findViewById(R.id.tv_clean_
                            cache_status);
        pm = getPackageManager();
        scanPackages();
    }
    //扫描出带有缓存的应用程序
    private void scanPackages() {
        //开启一个异步任务扫描带有缓存的应用程序
        new AsyncTask<Void, Integer, Void>() {
            //存储手机中所有已安装的应用程序的包信息
            List<PackageInfo> packinfos;
            @Override
            protected Void doInBackground(Void... params) {
                int i = 0;
                for (PackageInfo info : packinfos) {
                    //获取到应用程序的包名信息
                    String packname = info.packageName;
                    getSize(pm, packname);
```

```java
                i++;
                try {
                    Thread.sleep(100);
                } catch (InterruptedException e) {
                    //TODO Auto-generated catch block
                    e.printStackTrace();
                }
                publishProgress(i);
            }
            return null;
        }
        @Override
        protected void onPreExecute() {
            cachePagenames = new ArrayList<String>();
            cacheinfo = new HashMap<String, Long>();
            packinfos = pm.getInstalledPackages(0);
            pd.setMax(packinfos.size());
            tv_clean_cache_status.setText("开始扫描...");
            super.onPreExecute();
        }
        @Override
        protected void onPostExecute(Void result) {
            tv_clean_cache_status.setText("扫描完毕..." + "发现有"
                    + cachePagenames.size() + "个缓存信息");
            for (final String packname : cachePagenames) {
                //获取这些应用程序的图标、名称展现在界面上
                View child = View.inflate(getApplicationContext(),
                        R.layout.cache_item, null);
                //为child注册一个监听器
                child.setOnClickListener(new OnClickListener() {
                    //单击child时响应的单击事件
                    @Override
                    public void onClick(View v) {
                        //判断SDK的版本号
                        if (Build.VERSION.SDK_INT >= 9) {
//跳转至"清理缓存"的界面（可以通过：设置→应用程序→单击任意应用程序后的界面）
                            Intent intent = new Intent();
intent.setAction("android.settings.APPLICATION_DETAILS_SETTINGS");
                            intent.addCategory(Intent.CATEGORY_DEFAULT);
                            intent.setData(Uri.parse("package:" +
                                    packname));
                            startActivity(intent);
                        } else {
                            Intent intent = new Intent();
intent.setAction("android.intent.action.VIEW");
intent.addCategory(Intent.CATEGORY_DEFAULT);
```

第 5 章 其他模块的设计

```java
                intent.addCategory("android.intent.category.VOICE_LAUNCH");
                            intent.putExtra("pkg", packname);
                            startActivity(intent);
                        }
                    }
                });
                //为 child 中的控件设置数据
                ImageView iv_icon = (ImageView) child
                        .findViewById(R.id.iv_cache_icon);
            iv_icon.setImageDrawable(getApplicationIcon(packname));
                TextView tv_name = (TextView) child
                        .findViewById(R.id.tv_cache_name);
                tv_name.setText(getApplicationName(packname));
                TextView tv_size = (TextView) child
                        .findViewById(R.id.tv_cache_size);
                tv_size.setText("缓存大小 :"
                        + Formatter.formatFileSize
                                        (getApplicationContext(),
                                    cacheinfo.get(packname)));
                //将 child 添加到 ll_clean 控件上
                ll_clean.addView(child);
            }
            super.onPostExecute(result);
        }
    @Override
        protected void onProgressUpdate(Integer... values) {
            pd.setProgress(values[0]);
            tv_clean_cache_status.setText("正在扫描" + values[0] + "条目");
            super.onProgressUpdate(values);
        }
    }.execute();
}
//通过反射的方式调用 packageManager 中的方法
private void getSize(PackageManager pm, String packname) {
    try {
        //获取到 getPackageSizeInfo。调用 getPackageSizeInfo 方法需要在清
            单文件中配置权限信息: <uses-permission
        // android:name="android.permission.GET_PACKAGE_SIZE"/>
        Method method = pm.getClass().getDeclaredMethod(
                "getPackageSizeInfo",
                new Class[] { String.class, IPackageStatsObserver.class });
        //执行 getPackageSizeInfo 方法
        method.invoke(pm,
                new Object[] { packname, new MyObersver(packname) });
    } catch (Exception e) {
```

```java
            e.printStackTrace();
        }
    }
    //执行packageManager中的getPackageSizeInfo方法时需要传入
      IPackageStatsObserver.Stub接口,该接口通过aidl调用
    private class MyObersver extends IPackageStatsObserver.Stub {
        private String packname;
        public MyObersver(String packname) {
            this.packname = packname;
        }
        @Override
        public void onGetStatsCompleted(PackageStats pStats, boolean
                                    succeeded)
                throws RemoteException {
            //以下是根据ApplicationsState代码中的SizeInfo对象中定义的
            //缓存大小
            long cacheSize = pStats.cacheSize;
            //代码大小
            long codeSize = pStats.codeSize;
            //数据的大小
            long dataSize = pStats.dataSize;
            //判断这个包名对应的应用程序是否有缓存,如果有,则存入到集合中
            if (cacheSize > 0) {
                cachePagenames.add(packname);
                cacheinfo.put(packname, cacheSize);
            }
        }
    }
    //获取到应用程序的名称
    private String getApplicationName(String packname) {
        try {
            PackageInfo packinfo = pm.getPackageInfo(packname, 0);
            return packinfo.applicationInfo.loadLabel(pm).toString();
        } catch (Exception e) {
            e.printStackTrace();
            return packname;
        }
    }
    //获取到应用程序的图标
    private Drawable getApplicationIcon(String packname) {
        try {
            PackageInfo packinfo = pm.getPackageInfo(packname, 0);
            return packinfo.applicationInfo.loadIcon(pm);
        } catch (Exception e) {
            e.printStackTrace();
```

```
            return getResources().getDrawable(R.drawable.ic_launcher);
        }
    }
}
```

代码解析：

（1）new AsyncTask<Void, Integer, Void>()

开启一个异步任务。该异步任务的内部封装了 Handler+Message，开启了一个子线程执行任务。三种泛型类型分别代表"启动任务执行的输入参数"、"后台任务执行的进度"、"后台计算结果的类型"。在特定场合下，并不是所有类型都被使用，如果没有被使用，可以用 java.lang.Void 类型代替。异步任务的执行包含以下几个步骤：

① execute(Params... params)，执行一个异步任务，需要在代码中调用此方法，来触发异步任务的执行。

② onPreExecute()，在 execute(Params... params)被调用后立即执行，一般用于在执行后台任务前对 UI 做一些标记。

③ doInBackground(Params... params)，在 onPreExecute()完成后立即执行，用于执行较为费时的操作，此方法将接收输入参数和返回计算结果。在执行过程中可以调用 publishProgress(Progress... values)来更新进度信息。

④ onProgressUpdate(Progress... values)，在调用 publishProgress(Progress... values)时，此方法会被执行，直接将进度信息更新到 UI 组件上。

⑤ onPostExecute(Result result)，当后台操作结束时，此方法将会被调用。计算结果将作为参数传递到此方法中，直接将结果显示到 UI 组件上。

在使用的时候，还需要注意以下几点：

➢ 异步任务的实例必须在 UI 线程中创建；

➢ execute(Params... params)方法必须在 UI 线程中被调用；

➢ 不要手动调用 onPreExecute()、doInBackground(Params... params)、onProgressUpdate(Progress... values)、onPostExecute(Result result)方法；

➢ 不能在 doInBackground(Params... params)中更改 UI 组件的信息；

➢ 一个任务实例只能执行一次，如果执行第二次将会抛出异常。

（2）Method method = pm.getClass().getDeclaredMethod(
                    "getPackageSizeInfo",
                    new Class[] { String.class, IPackageStatsObserver.class })

通过反射的方式获取到 PackageManager 中的 getPackageSizeInfo 方法。getDeclaredMethod 中的第一个参数：要获取的方法名称；第二个参数：要获取的方法中的参数。

（3）method.invoke(pm,new Object[] { packname, new MyObersver(packname) })

执行 getPackageSizeInfo 方法，执行该方法需要在清单文件中加入权限：<uses-permission android:name="android.permission.GET_PACKAGE_SIZE"/>。其中，MyObersver 继承了 IPackageStatsObserver.Stub 接口，该接口是通过 aidl 的方式调用的。在 src 目录下新建一个包"android.content.pm"，在该包中需要拷入两个 aidl 文件：

第一个 aidl 文件：IPackageStatsObserver.aidl 文件内容如下：

```
/*
**
** Copyright 2007, The Android Open Source Project
**
** Licensed under the Apache License, Version 2.0 (the "License");
** you may not use this file except in compliance with the License.
** You may obtain a copy of the License at
**
**     http://www.apache.org/licenses/LICENSE-2.0
**
** Unless required by applicable law or agreed to in writing, software
** distributed under the License is distributed on an "AS IS" BASIS,
** WITHOUT WARRANTIES OR CONDITIONS OF ANY KIND, either express or implied.
** See the License for the specific language governing permissions and
** limitations under the License.
*/
package android.content.pm;
import android.content.pm.PackageStats;
/**
 * API for package data change related callbacks from the Package Manager.
 * Some usage scenarios include deletion of cache directory, generate
 * statistics related to code, data, cache usage(TODO)
 * {@hide}
 */
oneway interface IPackageStatsObserver {
    void onGetStatsCompleted(in PackageStats pStats, boolean succeeded);
}
```

第二个 aidl 文件：PackageStats.aidl 内容如下：

```
/* //device/java/android/android/view/WindowManager.aidl
**
** Copyright 2007, The Android Open Source Project
**
** Licensed under the Apache License, Version 2.0 (the "License");
** you may not use this file except in compliance with the License.
** You may obtain a copy of the License at
**
**     http://www.apache.org/licenses/LICENSE-2.0
**
```

```
  ** Unless required by applicable law or agreed to in writing, software
  ** distributed under the License is distributed on an "AS IS" BASIS,
  ** WITHOUT WARRANTIES OR CONDITIONS OF ANY KIND, either express or implied.
  ** See the License for the specific language governing permissions and
  ** limitations under the License.
  */
package android.content.pm;
parcelable PackageStats;
```

(4) if (Build.VERSION.SDK_INT >= 9)

判断 SDK 的版本号。因为 SDK 的版本号在升级到 9 时，将手机中的"设置"对应的意图修改了。

CleanCacheActivity 中的布局文件 ceanl_ce.xml 文件内容如下：

```
<LinearLayout xmlns:android="http://schemas.android.com/apk/res/android"
    xmlns:tools="http://schemas.android.com/tools"
    android:layout_width="fill_parent"
    android:layout_height="fill_parent"
    android:orientation="vertical" >
    <LinearLayout
        android:layout_width="fill_parent"
        android:layout_height="40dp"
        android:background="#ff6cbd45" >
        <TextView
            android:layout_width="fill_parent"
            android:layout_height="fill_parent"
            android:id="@+id/iv_main_title"
            android:gravity="center"
            android:text="缓 存 清 理"
            android:textColor="#ffffff"
            android:textSize="18sp"
            />
    </LinearLayout>
    <ProgressBar
        android:id="@+id/progressBar1"
        style="?android:attr/progressBarStyleHorizontal"
        android:layout_width="fill_parent"
        android:layout_height="wrap_content"
        ></ProgressBar>
    <TextView
        style="@style/text_content"
        android:text="正在扫描...."
         android:id="@+id/tv_clean_cache_status"
        />
    <ScrollView
        android:layout_height="fill_parent"
```

```xml
        android:layout_width="fill_parent"
        android:orientation="vertical"
        >
    <LinearLayout
        android:layout_width="fill_parent"
        android:layout_height="fill_parent"
        android:id="@+id/ll_clean_cache_cont"
        android:orientation="vertical"
        >
    </LinearLayout>
    </ScrollView>
      </LinearLayout>
```

文件解析：

（1）</ProgressBar>控件中引用的样式：style="?android:attr/progressBarStyleHorizontal"为系统自带的样式。

（2）</TextView>中引用的样式 style="@style/text_content"位于 res/values 目录下的 style.xml 文件中，style.xml 最终的文件内容如下：

```xml
<?xml version="1.0" encoding="utf-8"?>
<resources>
    <style name="text_title_style">
        <item name="android:layout_width">match_parent</item>
        <item name="android:layout_height">wrap_content</item>
        <item name="android:textColor">#66ff00</item>
        <item name="android:textSize">28sp</item>
    </style>
    <style name="image_divideline_style">
        <item name="android:layout_width">fill_parent</item>
        <item name="android:layout_height">1dip</item>
        <item name="android:layout_marginTop">5dip</item>
        <item name="android:background">@drawable/devide_line</item>
    </style>
    <style name="image_start_style">
        <item name="android:layout_width">wrap_content</item>
        <item name="android:layout_height">wrap_content</item>
        <item name="android:src">@android:drawable/star_big_on</item>
    </style>
    <style name="image_online_style">
        <item name="android:layout_width">wrap_content</item>
        <item name="android:layout_height">wrap_content</item>
        <item name="android:paddingLeft">3dip</item>
        <item name="android:src">@android:drawable/presence_online</item>
    </style>
    <style name="image_offline_style">
        <item name="android:paddingLeft">3dip</item>
```

```xml
        <item name="android:layout_width">wrap_content</item>
        <item name="android:layout_height">wrap_content</item>
        <item name="android:src">@android:drawable/presence_invisible</item>
    </style>
    <style name="image_logo_style">
        <item name="android:layout_width">fill_parent</item>
        <item name="android:layout_height">fill_parent</item>
        <item name="android:scaleType">center</item>
    </style>
    <style name="button_next_style">
        <item name="android:layout_width">wrap_content</item>
        <item name="android:layout_height">wrap_content</item>
        <item name="android:layout_alignParentBottom">true</item>
        <item name="android:layout_alignParentRight">true</item>
        <item name="android:text">下一步</item>
        <item name="android:drawableRight">@drawable/next</item>
        <item name="android:onClick">next</item>
    </style>
    <style name="button_pre_style">
        <item name="android:layout_width">wrap_content</item>
        <item name="android:layout_height">wrap_content</item>
        <item name="android:layout_alignParentBottom">true</item>
        <item name="android:layout_alignParentLeft">true</item>
        <item name="android:text">上一步</item>
        <item name="android:onClick">pre</item>
        <item name="android:drawableLeft">@drawable/previous</item>
    </style>
    <style name="text_content_style" parent="@style/text_title_style">
        <item name="android:textSize">20sp</item>
    </style>
    <style name="my_pb_style" parent="@android:style/Widget.ProgressBar">
        <item name="android:indeterminateDrawable">@drawable/my_pb_bg</item>
    </style>
    <style name="text_title_style2">
        <item name="android:layout_width">fill_parent</item>
        <item name="android:layout_height">fill_parent</item>
        <item name="android:gravity">left</item>
        <item name="android:textColor">#ffffff</item>
        <item name="android:textSize">18sp</item>
    </style>
     <style name="text_content" parent="@style/text_title_style2">
        <item name="android:layout_height">wrap_content</item>
        <item name="android:textColor">#ffffff</item>
        <item name="android:textSize">16sp</item>
    </style>
</resources>
```

（3）最后一个</LinearLayout>控件是用于添加扫描到带有缓存应用的 View（该 View 中主要包含应用的图标、名称、缓存大小信息），该 View 对应的布局文件 cache_item.xml 文件内容如下：

```xml
<?xml version="1.0" encoding="utf-8"?>
<RelativeLayout xmlns:android="http://schemas.android.com/apk/res/android"
    android:layout_width="match_parent"
    android:layout_height="wrap_content" >
    <ImageView
        android:id="@+id/iv_cache_icon"
        android:layout_width="30dp"
        android:layout_height="30dp"
        android:src="@drawable/ic_launcher" />
    <TextView
        android:id="@+id/tv_cache_name"
        android:layout_width="wrap_content"
        android:layout_height="wrap_content"
        android:layout_toRightOf="@id/iv_cache_icon"
        android:text="包名" />
    <TextView
        android:id="@+id/tv_cache_size"
        android:layout_width="wrap_content"
        android:layout_height="wrap_content"
        android:layout_below="@id/tv_cache_name"
        android:layout_toRightOf="@id/iv_cache_icon"
        android:text="缓存大小"
        android:textColor="#66000000" />
    <ImageView
        android:layout_width="30dp"
        android:layout_height="30dp"
        android:layout_alignParentRight="true"
        android:src="@drawable/block_delete_btn" />
</RelativeLayout>
```

最后，在主界面中为 CleanCacheActivity 设置单击事件，MainActivity.java 中的最终业务代码如下：

```java
package com.guoshisp.mobilesafe;
import android.app.Activity;
import android.content.Intent;
import android.os.Bundle;
import android.view.View;
import android.widget.AdapterView;
import android.widget.AdapterView.OnItemClickListener;
import android.widget.GridView;
import com.guoshisp.mobilesafe.adapter.MainAdapter;
public class MainActivity extends Activity {
```

```java
//显示主界面中九大模块的GridView
private GridView gv_main;
@Override
protected void onCreate(Bundle savedInstanceState) {
    super.onCreate(savedInstanceState);
    setContentView(R.layout.main);
    gv_main = (GridView) findViewById(R.id.gv_main);
//为gv_main对象设置一个适配器，该适配器的作用是用于为每个item填充对应的数据
    gv_main.setAdapter(new MainAdapter(this));
    //为GridView对象中的item设置单击时的监听事件
    gv_main.setOnItemClickListener(new OnItemClickListener() {
        //参数一：item的父控件，也就是GridView；参数二：当前单击的item；参
        //    数三：当前单击的item在GridView中的位置
        //参数四：id的值为单击了GridView的哪一项对应的数值，单击了GridView
        //    第9项，那id就等于8
        public void onItemClick(AdapterView<?> parent, View view,
                int position, long id) {
            switch (position) {
            case 0: //手机防盗
                //跳转到"手机防盗"对应的Activity界面
                Intent lostprotectedIntent = new Intent(MainActivity.
                        this,LostProtectedActivity.class);
                startActivity(lostprotectedIntent);
                break;
            case 1: //通信卫士
                Intent callSmsIntent = new Intent(MainActivity.this,
                        CallSmsSafeActivity.class);
                startActivity(callSmsIntent);
                break;
            case 2: //程序管理
                Intent appManagerIntent = new Intent(MainActivity.this,
                        AppManagerActivity.class);
                startActivity(appManagerIntent);
            case 3: //进程管理
                Intent taskManagerIntent = new Intent(MainActivity.this,
                        TaskManagerActivity.class);
                startActivity(taskManagerIntent);
                break;
            case 4: //流量管理
                Intent trafficInfoIntent = new Intent(MainActivity.this,
                        TrafficInfoActivity.class);
                startActivity(trafficInfoIntent);
                break;
            case 5: //手机杀毒
                Intent antiVirusIntent = new Intent(MainActivity.this,
                        AntiVirusActivity.class);
```

```
            startActivity(antiVirusIntent);
            break;
        case 6: //系统优化
            Intent cleanCacheIntent = new Intent(MainActivity.this,
                            CleanCacheActivity.class);
            startActivity(cleanCacheIntent);
            break;
        case 7://高级工具
            Intent atoolsIntent = new Intent(MainActivity.this,
                            AtoolsActivity.class);
            startActivity(atoolsIntent);
            break;
        case 8://设置中心
            //跳转到"设置中心"对应的Activity界面
            Intent settingIntent = new Intent(MainActivity.this,
                            SettingCenterActivity.class);
            startActivity(settingIntent);
            break;
        }
    }
});
}
```

测试运行：在清单文件中配置好组件、权限信息后，单击"系统优化"后的界面如图5-40所示。运行完成后的界面如图5-41所示。单击任意一个带有缓存信息的应用后的界面如图5-42所示。单击图5-42中的"清除缓存"后的界面如图5-43所示。

图 5-40   图 5-41

　　　　图 5-42　　　　　　　　　　图 5-43

 需要掌握的知识点小结

（1）如何查看系统隐藏的 API。

（2）使用反射技术来调用 PackageManager 中的方法。

（3）自定义样式。

（4）使用 AsyncTask 处理异步任务。

作者介绍

王家林：

Android 架构师、高级工程师、咨询顾问、培训专家，通晓 Android、HTML5、Hadoop，致力于 Android 和 HTML5 软、硬、云整合。

HTML5 技术领域的最早实践者（2009 年）之一，成功为多个机构实现多款自定义 HTML5 浏览器，参与某知名的 HTML5 浏览器研发。

成功对包括三星、摩托罗拉、华为等世界 500 强企业实施 Android 底层移植、框架修改、应用开发等培训；成功对平安保险、英特尔等实施 HTML5 培训。项目案例包括 Android 移植工作、Android 上特定硬件的垂直整合、编写 Java 虚拟机、Android 框架修改、Android 手机卫士开发、Android 娱乐多媒体软件开发、大型 B2C 电子商务网站开发、SNS 网站开发等。

致力于提供 Android&HTML5&Hadoop 全方位一站式的培训和咨询解决方案，最具代表性的课程有：

- Android 系统完整训练系列：开发搭载 Android 系统的产品；
- Android 框架深入浅出系列：HAL&Framework&Native Service &Android，Service 架构设计与实战开发；
- Android 软、硬整合框架精髓实战；
- AndroidFramework 系统整合与维护；
- 面向 Web Cloud 的 HTML5 App 开发实战系列：Browser&HTML5&CSS3&PhoneGap& jQueryMobile& WebSocket&Node.js；
- HTML5 端云整合：智能端应用与云端服务整合开发实战；
- Android 平台开发最佳实践课程。
- Android 高级 UI 技术最佳实践课程。

机构介绍

王家林所创立并领导的国士工作室是一支专注于 Android、HTML5、Hadoop 企业级实战开发的技术团队，致力于做国内优秀的 Android、HTML5、Hadoop 程序开发机构，提供完善的 Android、HTML5、Hadoop 企业级开发培训和咨询服务。国士工作室在基于实务的基础上推出了 Android 与 HTML5 的一站式培训课程。

Android 课程如下：

CourseID	课 程 名 称	开 课 类 型	课 程 时 长
AF101	Android 系统完整训练：开发搭载 Android 系统的产品	公开课/企业内训	24 小时
AF102	Android 框架深入浅出：HAL&Framework&Native Service &Android Service 架构设计与实战开发	公开课/企业内训	18 小时
AF103	Android 软、硬整合框架精髓实战	公开课/企业内训	18 小时
AF104	Android Framework 系统整合与维护	公开课/企业内训	12 小时
AF105	Android 4.x porting：移植技术与实战训练	公开课/企业内训	12 小时
AF106	Android Binder IPC Subsystem	企业内训	12 小时
AF107	Android UI & View Subsystem 架构与设计解析	企业内训	12 小时
AF108	Android ActivityManager 架构与设计解析	企业内训	12 小时
AF109	Android WindowManager 架构与设计解析	企业内训	12 小时
AF110	Application Launching & Launcher Design	企业内训	12 小时
AD201	Android 平台应用开发最佳实践	公开课/企业内训	24 小时
AD202	精通移动互联网下 Android 应用程序开发实战	公开课/企业内训	18 小时
AD203	精通 Android 高级 UI 技术最佳实践	公开课/企业内训	12 小时
AT301	云时代 Android 应用测试最佳实践	公开课/企业内训	18 小时
AT302	Android 系统测试最佳实践	公开课/企业内训	18 小时

HTML5 课程如下：

CourseID	课 程 名 称	开 课 类 型	课 程 时 长
WB101	面向 Web Cloud 的 HTML5 App 开发实战：Browser&HTML5&CSS3& PhoneGap&jQuery Mobile& WebSocket&Node.js	公开课/企业内训	12 小时
WB102	HTML5 端云整合：智能端应用与云端服务整合开发实战	公开课/企业内训	18 小时
WB103	Node.js 与云端服务开发	企业内训	12 小时
WB104	云端数据库入门：NoSQL 与 Open API 实战	企业内训	12 小时

Hadoop 课程如下：

CourseID	课 程 名 称	开 课 类 型	课 程 时 长
CH101	云计算实战：Hadoop 开发全程代码实战（面向软件工程师、数据库工程师、网络后台开发人员等）	公开课/企业内训	12 小时
CH102	云计算实战：Hadoop 数据库管理员实战（面向数据库管理员、系统管理员等）	公开课/企业内训	18 小时
CH103	Hadoop 深入浅出开发实战	企业内训	24 小时
CH104	王家林的云计算实战：Hadoop 大数据处理之生态系统和成功案例（面向 CIO、CTO、DBA、架构师等）	企业内训	6 小时

Linux 课程如下：

CourseID	课 程 名 称	开 课 类 型	课 程 时 长
LD101	嵌入式 Linux 架构和开发实战	公开课/企业内训	12 小时
LD102	Linux 驱动开发与案例实战	公开课/企业内训	18 小时
LT201	实战测试驱动开发在嵌入式系统中的应用	公开课/企业内训	12 小时